Cliffs Quick Review

P9-CMQ-394

Chemistry

by
Harold D. Nathan, Ph.D.

Series Editor
Jerry Bobrow, Ph.D.

INCORPORATED
LINCOLN, NEBRASKA   68501

Cliffs Quick Review Chemistry

The Cliffs Notes logo, the names "Cliffs," "Cliffs Notes," and "Cliffs Quick Review," and the black and yellow diagonal stripe cover design are all registered trademarks belonging to Cliffs Notes, Inc., and may not be used in whole or in part without written permission.

FIRST EDITION

ISBN 0-8220-5318-7

CONTENTS

CONTENTS

CONTENTS

INTRODUCTION

Chemistry is the science that studies the substances that constitute all matter. It is a systematic interpretation of the properties of such substances, or of higher level substances formed by combination, or of lower level substances resulting from decomposition.

Discovery and Similarity

The modern science of chemistry began during the eighteenth century when several brilliant "natural philosophers" classified the products of decomposition into a small number of fundamental substances. One famous instance is the discovery—in 1774 by the Englishman Joseph Priestley—that when the red powder called mercuric oxide is heated, it decomposes to liquid metal mercury and a colorless gas capable of supporting combustion. (This gas was later named oxygen.) Most substances can similarly be decomposed into several simpler substances either by heat or an electrical current; however, the most fundamental substances could not be broken down further, even with extraordinary temperature or electric voltage. Those basic building blocks of all other substances came to be called the chemical **elements**.

When the French chemist Antoine Lavoisier published his famous list of elements in 1789, there were only 33 of them, and several of those were erroneous. By 1930, the diligent labors of thousands of chemists had increased the tally of naturally occurring chemical elements to 90. More recently, physicists in high-energy laboratories have been able to create about 20 highly radioactive, unstable elements that do not exist naturally on Earth, although they are probably produced in the hot cores of some stars.

The number of chemical elements has now reached 109. Fortunately for students, only about 40 are of importance to basic chemistry. Please take a reconnaissance glance at the Periodic Table of the Chemical Elements (pages 10-11) and find calcium, element number 20. You will need to be acquainted with the properties of the 20 simplest elements, up to calcium, plus another 20 of chemical significance that you will encounter in this book.

You will notice that the key concepts of chemistry are set in boldface on their first appearance in the text to alert you to their importance. These terms are used repeatedly here, and you cannot master chemistry without understanding them. Definitions of the key concepts are collected in Appendix A: Glossary of Chemical Terms, page 155, to which you should refer whenever necessary.

Most elements have chemical and physical properties that strongly resemble those of other elements. For example, helium, neon, argon, krypton, and xenon are all colorless gases that combine with other elements only under very special conditions; their lack of reactivity leads to the name **inert gases** (or noble gases) for this group of similar elements. By contrast, fluorine and chlorine are corrosive greenish gases that form salts when they readily combine with metals, hence the name **halogens** ("salt formers") for fluorine, chlorine, bromine, and iodine.

THREE GROUPS OF SIMILAR ELEMENTS

3		2
Li		**He**
Lithium		Helium
6.94		4.00

11	9	10
Na	**F**	**Ne**
Sodium	Fluorine	Neon
22.99	19.00	20.18

19	17	18
K	**Cl**	**Ar**
Potassium	Chlorine	Argon
39.10	35.45	39.95

37	35	36
Rb	**Br**	**Kr**
Rubidium	Bromine	Krypton
85.47	79.90	83.80

55	53	54
Cs	**I**	**Xe**
Cesium	Iodine	Xenon
132.91	126.90	131.30

| Alkali metals | The halogens | Inert gases |

■ Figure 1 ■

As a final example of a group of elements with similar properties, the metallic elements lithium, sodium, and potassium have such low densities that they float on water and are so highly reactive that they spontaneously burn by extracting oxygen from the water itself; these light metals form strong alkalis and are appropriately called the **alkali metals.** You should locate each of these columns of similar elements on the periodic table.

Similar elements also occur in the same natural environment. For instance, the halogens are markedly concentrated in seawater. You are aware that the major salt in ocean brines is sodium chloride. The other halogens are extracted from seawater that has been further concentrated: bromine from salt beds formed by evaporation, and iodine from kelp growing in oceans.

The first, indispensable key to making sense out of the extensive system of facts and principles called chemistry is the rule that the behavior of an element or compound can be predicted from similar substances.

Atomic Weights

By the early nineteenth century, chemists were striving to organize their rudimentary knowledge of the chemical elements. It was known that differing weights of elements reacted to form compounds. As an example, they found that 3 grams of magnesium metal reacted with precisely 2 grams of oxygen to form magnesium oxide with no residual magnesium or oxygen. The same weight of oxygen, however, required 5 grams of calcium metal to react completely to calcium oxide. The following chart summarizes those relative combining weights:

RATIOS FROM EXPERIMENTS

Combining weights

5	calcium
3	magnesium
2	oxygen

Chemists gradually discovered that such relative weights in chemical reactions were fundamental characteristics of the elements. The English chemist John Dalton realized that all the combining weights then known were nearly whole-number multiples of the combining weight of the lightest element, hydrogen. In 1803, he proposed an atomic theory in which all other elements would be built from multiple hydrogen atoms. Consequently, he based his scale of **atomic weights** on hydrogen being equal to 1.

Although Dalton's theory was found to be unrealistically simple, he did compel chemists to adopt a standard scale of atomic weights. Because the combining weight of oxygen is approximately 16 times that of hydrogen, the earlier chart can be revised:

STANDARD SCALE

Atomic weights

40.08	calcium
24.31	magnesium
16.00	oxygen
1.01	hydrogen

The modern weights for calcium, magnesium, and oxygen are still nearly in the 5:3:2 ratios of the original table. Notice especially that the atomic weight of hydrogen is not precisely equal to 1 because the scale is now defined by the most common variety of carbon being exactly 12 atomic weight units. Dalton's bold conjecture that all the heavier elements have weights that are integral multiples of hydrogen is not strictly valid, but it is a good approximation that eventually led to the discovery of the particles composing the atoms.

The Periodic Table

In 1869, the Russian chemist Dmitri Mendeleev published his great systematization called the **periodic table**. He arranged all known chemical elements in order of their atomic weights and found that similar physical and chemical properties recurred every 7 elements for the lighter elements and every 17 elements for the heavier ones. (The inert gases had not been discovered at that time; the correct values for similar properties are 8 and 18.) The periodic table is based on atomic weights and similar properties. In each row, the atomic weights increase toward the right. Each column contains a group of elements with similar chemical behavior.

The modern periodic table is on pages 10-11. Notice that each box contains four data. Besides the element's name and symbol, the atomic weight is at the bottom, and an atomic number is at the top.

READING THE PERIODIC TABLE

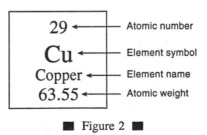

■ Figure 2 ■

In the preceding section, you reviewed the historical determination of atomic weights. You should examine the periodic table to satisfy yourself that almost all the elements are arranged in order of increasing atomic weight, but there are several exceptions: compare the atomic weights of cobalt (Co) to nickel (Ni) and also compare tellurium (Te) to iodine (I). In these cases, the element order is based on the rule that similar elements must be aligned vertically.

TELLURIUM AND IODINE

16	17
S	Cl
Sulfur	Chlorine
32.06	35.45
34	35
Se	Br
Selenium	Bromine
78.96	79.90
52	53
Te	I
Tellurium	Iodine
127.60	126.90

(Compare their atomic weights.)

■ Figure 3 ■

Let's discuss the proper placement of tellurium and iodine in the table, where Te has the heavier atomic weight. The chemical properties of tellurium are like those of selenium because both are semimetallic elements that form compounds like those of sulfur. Iodine resembles bromine because they are nonmetallic halogens that form compounds like those of chlorine. Therefore, the order in the table cannot be based solely on atomic weight.

The **atomic number**, which appears above each element symbol, represents the meaningful order in the periodic table. Whenever an element is referred to by an integer, that means the atomic number, not the atomic weight. Thus, element 27 is cobalt (atomic number is 27), not aluminum (atomic weight is 27). In a later section (page 21), the two terms will be more carefully defined; for now, simply bear in mind the distinction between atomic number and atomic weight.

The periodic table displays the pattern of properties of the elements. The lightest are at the top of the chart; the atomic weights increase toward the bottom of the chart. The elements to the upper right, above a diagonal line from aluminum (13) to polonium (84), are **nonmetals**, about half of which exist as gases under normal laboratory conditions. All the elements in the middle and left of the table are **metals**, except gaseous hydrogen (1). Most of the metals are shiny, deformable solids, but mercury has such a low melting point that it is a liquid at room temperature. All the metals have high conductivities for heat and electricity. *Many simple chemical compounds are formed from a metal reacting with a nonmetal.* In the periodic table, elements in columns have similar properties, and elements so related (like sulfur, selenium, and tellurium) are said to be **congeners**.

PERIODIC TABLE OF THE CHEMICAL ELEMENTS

1 H Hydrogen 1.01								
3 Li Lithium 6.94	4 Be Beryllium 9.01							
11 Na Sodium 22.99	12 Mg Magnesium 24.31							
19 K Potassium 39.10	20 Ca Calcium 40.08	21 Sc Scandium 44.96	22 Ti Titanium 47.90	23 V Vanadium 50.94	24 Cr Chromium 52.00	25 Mn Manganese 54.94	26 Fe Iron 55.85	27 Co Cobalt 58.93
37 Rb Rubidium 85.47	38 Sr Strontium 87.62	39 Y Yttrium 88.91	40 Zr Zirconium 91.22	41 Nb Niobium 92.91	42 Mo Molybdenum 95.94	43 Tc Technetium (99)	44 Ru Ruthenium 101.07	45 Rh Rhodium 102.91
55 Cs Cesium 132.91	56 Ba Barium 137.34	57 La Lanthanum 138.91	72 Hf Hafnium 178.49	73 Ta Tantalum 180.95	74 W Tungsten 183.85	75 Re Rhenium 186.21	76 Os Osmium 190.2	77 Ir Iridium 192.22
87 Fr Francium (223)	88 Ra Radium (226)	89 Ac Actinium (227)	104 (261)	105 (262)	106 Names not yet established for these elements (263)	107 (262)	108 (265)	109 (266)

	58 Ce Cerium 140.12	59 Pr Praseodymium 140.91	60 Nd Neodymium 144.24	61 Pm Promethium (147)	62 Sm Samarium 150.35	63 Eu Europium 151.96
Lanthanides						
Actinides	90 Th Thorium (232)	91 Pa Protactinium (231)	92 U Uranium (238)	93 Np Neptunium (237)	94 Pu Plutonium (242)	95 Am Americium (243)

								2 **He** Helium 4.00
			5 **B** Boron 10.81	6 **C** Carbon 12.01	7 **N** Nitrogen 14.01	8 **O** Oxygen 16.00	9 **F** Fluorine 19.00	10 **Ne** Neon 20.18
			13 **Al** Aluminum 26.98	14 **Si** Silicon 28.09	15 **P** Phosphorus 30.97	16 **S** Sulfur 32.06	17 **Cl** Chlorine 35.45	18 **Ar** Argon 39.95
28 **Ni** Nickel 58.71	29 **Cu** Copper 63.55	30 **Zn** Zinc 65.38	31 **Ga** Gallium 69.72	32 **Ge** Germanium 72.59	33 **As** Arsenic 74.92	34 **Se** Selenium 78.96	35 **Br** Bromine 79.90	36 **Kr** Krypton 83.80
46 **Pd** Palladium 106.42	47 **Ag** Silver 107.87	48 **Cd** Cadmium 112.41	49 **In** Indium 114.82	50 **Sn** Tin 118.69	51 **Sb** Antimony 121.75	52 **Te** Tellurium 127.60	53 **I** Iodine 126.90	54 **Xe** Xenon 131.30
78 **Pt** Platinum 195.09	79 **Au** Gold 196.97	80 **Hg** Mercury 200.59	81 **Tl** Thallium 204.37	82 **Pb** Lead 207.19	83 **Bi** Bismuth 208.98	84 **Po** Polonium (210)	85 **At** Astatine (210)	86 **Rn** Radon (222)

64 **Gd** Gadolinium 157.25	65 **Tb** Terbium 158.93	66 **Dy** Dysprosium 162.50	67 **Ho** Holmium 164.93	68 **Er** Erbium 167.26	69 **Tm** Thulium 168.93	70 **Yb** Ytterbium 173.04	71 **Lu** Lutetium 174.97
96 **Cm** Curium (247)	97 **Bk** Berkelium (247)	98 **Cf** Californium (251)	99 **Es** Einsteinium (254)	100 **Fm** Fermium (257)	101 **Md** Mendelevium (258)	102 **No** Nobelium (259)	103 **Lr** Lawrencium (260)

Chemical Compounds

The whole-number ratios of combining weights in chemical reactions would be readily explained if the basic unit of all elements were the **atom** because that word originally meant an indivisible particle. Therefore, one atom of carbon could react with either one atom of oxygen,

$$C \quad + \quad O \quad \longrightarrow \quad CO$$

carbon	oxygen	carbon
atom	atom	monoxide
		molecule

or two atoms of oxygen,

$$C \quad + \quad O_2 \quad \longrightarrow \quad CO_2$$

carbon	oxygen	carbon
atom	molecule	dioxide
		molecule

but not with, say, 1½ oxygen atoms. Notice that in writing chemical formulas the subscripts (the 2 in O_2 and in CO_2) represent the relative number of atoms of each element.

A substance containing atoms of more than one element is called a **compound**. In the preceding example showing the combination of carbon and oxygen to carbon dioxide, both carbon and oxygen are elements, while the carbon dioxide is a compound. When several atoms are so tightly bonded together that they physically behave as a unit, that group is called a **molecule**. *Either elements or compounds can be molecular.* In the carbon dioxide reaction, the O_2 molecule contains 2 atoms of oxygen, and the CO_2 molecule has 3 atoms, 1 carbon and 2 oxygen.

ATOMS AND MOLECULES

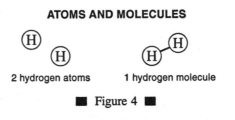

2 hydrogen atoms   1 hydrogen molecule

■ Figure 4 ■

Five major compounds are built solely of nitrogen and oxygen, as shown in the following chart. Of the five oxides of nitrogen, nitrous oxide is richest in nitrogen, and dinitrogen pentoxide is richest in oxygen.

THE OXIDES OF NITROGEN

Compound name	Chemical formula	Atomic ratio N:O
Nitrous oxide	N_2O	2:1
Nitric oxide	NO	1:1
Dinitrogen trioxide	N_2O_3	2:3
Nitrogen dioxide	NO_2	1:2
Dinitrogen pentoxide	N_2O_5	2:5

Stoichiometry

The atomic ratios in each compound are also the relative number of atomic weight units of its elements. The first example is nitrous oxide (N_2O).

NITROUS OXIDE

Chemical element	Atomic ratio	Atomic weight	Relative weight	Weight percent
Nitrogen	2	14.01	28.02	63.65
Oxygen	1	16.00	16.00	36.35

The values in the fourth column were obtained by multiplying the atomic ratios and atomic weights. You can see that a sample of N_2O weighing 44.02 grams would contain 28.02 g of nitrogen and 16.00 g of oxygen. The weight percent of each element is calculated from its relative weight divided by the sum of the relative weights. Right now, why don't you try to calculate the values in the fifth column from those in the fourth column? Chemical compounds with integral atomic ratios, like nitrous oxide, are described as **stoichiometric** compounds and permit many simple calculations.

The common oxide of aluminum will provide a second example, but this time, let's begin with the weight percent and deduce the atomic ratio. Careful laboratory analysis of aluminum oxide would find it to be approximately 53% aluminum and 47% oxygen by weight, as shown in the second column of the next chart.

ALUMINUM OXIDE

Chemical element	Weight percent	Atomic weight	Relative weight	Atomic ratio
Aluminum	52.92	26.98	1.961	2
Oxygen	47.08	16.00	2.943	3

The relative weights in the fourth column are obtained by dividing the weight percentages by the atomic weights from the periodic table.

$$\text{aluminum} \qquad \frac{52.92}{26.98} = 1.961$$

$$\text{oxygen} \qquad \frac{47.08}{16.00} = 2.943$$

So, the ratio of oxygen to aluminum in relative atomic weights is

$$\frac{\text{oxygen}}{\text{aluminum}} = \frac{2.942}{1.961} = 1.50$$

If the aluminum oxide is a stoichiometric compound, with whole-number ratios of the constituent elements, the quotient calculated above must be translated into a ratio of integers:

$$1.50 = \frac{3}{2} = \frac{\text{oxygen}}{\text{aluminum}}$$

That ratio of integers implies that the formula for aluminum oxide is Al_2O_3. In simple compounds, the metallic elements are written before the nonmetals.

Most chemical compounds are stoichiometric, and you should be able to utilize the atomic weights and perform either calculation:

1. Weight percents from the chemical formula
2. Chemical formula from the weight percents

These calculations are so basic to the field that you should go back and carefully review the two examples in this section: the calculation of weight percents in N_2O and the inference of the formula for aluminum oxide. Then, you can practice stoichiometric calculations on the following pair of problems, which are answered and explained in Appendix B: Answers to Problems (page 161). Many such practice exercises have been included in this book so that you can determine whether you understand the major concepts of chemistry. It is well worth your time to study these examples and their explanations in Appendix B until you can do the calculations correctly.

Problem 1. Determine the weight percentages of the three elements in ammonium chloride, which has the formula NH_4Cl.

Problem 2. Infer the simplest chemical formula for potassium copper fluoride, which has this analysis by weight:

Potassium	35.91%
Copper	29.19%
Fluorine	34.90%

The Mole Unit

In laboratory practice, you work with much larger quantities of the elements than single atoms or single molecules. A convenient standard quantity is the **mole**, the amount of the substance equal in grams to the sum of the atomic weights. The word *mole* is a contraction from *molecular weight*. One mole of carbon dioxide, therefore, weighs 44.01 grams:

$$C \quad + \quad 2(O) \quad = \quad CO_2$$

carbon   oxygen   carbon dioxide

$$12.01 \quad + \quad 2(16.00) \quad = \quad 44.01$$

The mole is a convenient unit for expressing the relative amounts of substances in chemical reactions. The burning of carbon in oxygen would be written with the 2 oxygen atoms bonded in a single O_2 molecule:

$$C \quad + \quad O_2 \quad \longrightarrow \quad CO_2$$

1 mole	1 mole	1 mole
carbon	oxygen	carbon dioxide
(12.01 g)	(32.00 g)	(44.01 g)

The mole is the most common unit used to express the quantity of a chemical substance. For all solids, liquids, and gases, you can convert weight to moles (or moles to weight). For gases, you should memorize the following conversion of volume to moles (or moles to volume): at 0°C and 1 atmosphere pressure, *one mole of any gas occupies approximately 22.4 liters*. Therefore, the preceding reaction describing the oxidation of carbon means that 12 grams of carbon burned in 22.4 liters of oxygen yields 22.4 liters of carbon dioxide.

Many substances do not exist as molecules. For example, the atoms in most inorganic solids are in a three-dimensional structure where each atom is surrounded by a number of other atoms. In crystals of sodium chloride, there are no distinct pairs of Na and Cl atoms that could be called molecules because each sodium is surrounded by 6 chlorine atoms and each chlorine is surrounded by 6 sodium atoms. (See Figure 58 on page 145.)

For nonmolecular substances like sodium chloride, the use of the word *mole*, with its connotation of molecules, would be inappropriate. A comparable unit, the **gram formula weight**, is used; it is defined as the sum in grams of atomic weights in the chemical formula of the substance. For sodium chloride (NaCl), one gram formula weight is calculated as

$$22.99 \quad + \quad 35.45 \quad = \quad 58.44 \text{ grams}$$

22.99	35.45	
sodium	chlorine	

The mole unit and the gram formula unit are employed in similar calculations. For that reason, many chemists use the term *mole* to describe quantities of both molecular and nonmolecular substances.

Problem 3. How many moles are there in 1 kilogram of bromobenzene, C_6H_5Br?

Problem 4. What is the mass in grams of neon gas that occupies a volume of 5 liters at 0°C and atmospheric pressure?

Chemical Reactions

The standard representation of a chemical reaction shows an arrow pointing from the **reactants** toward the **products**:

$$Fe_2O_3(s) + 3CO(g) \longrightarrow 2Fe(s) + 3CO_2(g)$$

| iron oxide | carbon monoxide | metallic iron | carbon dioxide |

For most chemical reactions in this book, solids are labeled (s), liquids (l), and gases (g). The numerical coefficients in front of the chemical formulas express the relative moles of each compound or element. The preceding reaction can be interpreted in terms of either moles or weights.

INTERPRETATION OF A REACTION				
Quantity	Fe_2O_3	CO	Fe	CO_2
Moles	1	3	2	3
Weight of 1 mole	159.70	28.01	55.85	44.01
Total weights	159.70	84.03	111.70	132.03

Notice that the total weights of the reactants (243.73g) equal the total weights of the products. This principle is that of the conservation of matter.

It is *not* true, however, that volumes must be conserved in reactions involving gases. The complete combustion of carbon monoxide is a case in point:

$$2CO(g) \ + \ O_2(g) \longrightarrow 2CO_2(g)$$

carbon   oxygen   carbon
monoxide   dioxide

Because the reaction coefficients are proportional to relative volumes of each gas, 2 volumes of carbon dioxide and 1 volume of oxygen (a total of 3 volumes of reactants) combine to produce only 2 volumes of carbon dioxide. In gaseous reactions, the total volume of the products may be less than, or equal to, or greater than the total volume of the reactants.

Problem 5. How many liters of oxygen would be required for the complete oxidation of 1 gram of methane?

$$CH_4(g) \ + \ 2O_2(g) \longrightarrow CO_2(g) + 2H_2O(g)$$

methane   oxygen   carbon   water
                         dioxide   vapor

ATOMIC STRUCTURE

Subatomic Particles

The rather steady increase of atomic weights through the periodic table was explained when physicists managed to split atoms into three component particles.

PARTS OF AN ATOM			
Subatomic particle	Mass units	Electric charge	Atomic location
Proton	1.0073	+1	nucleus
Neutron	1.0087	0	nucleus
Electron	0.0005	−1	orbital

The exploration of atomic structure began in 1911 when Ernest Rutherford, a New Zealander who worked in Canada and England, discovered that atoms had a dense central **nucleus** that contained positively charged particles, which he named **protons**. It was soon established that each chemical element was characterized by a specific number of protons in each atom. A hydrogen atom has 1 proton, helium has 2, lithium has 3, and so forth through the periodic table. *The atomic number is the number of protons* for each element.

Except for the simplest hydrogen atom with a single proton as its entire nucleus, all atoms contain uncharged **neutrons** (the word means "electrically neutral") in addition to the characteristic protons. For most of the light elements, the numbers of protons and neutrons are nearly equal. The following chart shows the most common nucleus for each element with the atomic weight rounded to the nearest integer. You can see that *the atomic weights are the sum of the protons and neutrons* for each atom.

NUCLEAR STRUCTURE
OF THE LIGHTEST ELEMENTS

Element	Atomic number	Protons	Neutrons	Atomic weight
Hydrogen	1	1	0	1
Helium	2	2	2	4
Lithium	3	3	4	7
Beryllium	4	4	5	9
Boron	5	5	6	11
Carbon	6	6	6	12
Nitrogen	7	7	7	14
Oxygen	8	8	8	16

John Dalton's interpretation that whole-number atomic weights are multiple hydrogen atoms was premature, but near the truth. The series of elements of increasing atomic weights is generated by adding **nucleons,** the two types of particles comprising the nucleus.

Isotopes

Although most of the lighter elements have atomic weights that are nearly whole numbers, some elements were discovered to have atomic weights that could not be integral. Look at the atomic weights of the three lightest halogens and satisfy yourself that although the values for

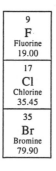

9
F
Fluorine
19.00

17
Cl
Chlorine
35.45

35
Br
Bromine
79.90

■ Figure 5 ■

fluorine and bromine might be whole numbers, that of chlorine is definitely intermediate. The interpretation of the curious weight of chlorine awaited the discovery of the neutron in 1932. Although all chlorine atoms have 17 protons, different **isotopes** of the element have different numbers of neutrons. In the chart, the masses of the chlorine isotopes are denoted by superscripts to the upper left of the chemical symbol.

ISOTOPES OF CHLORINE

Isotope	Protons	Neutrons	Atomic mass	Natural abundance
^{35}Cl	17	18	34.97	76%
^{37}Cl	17	20	36.97	24%

The nonintegral atomic weight for naturally occurring chlorine is seen to be the weighted average of its two major isotopes:

$$0.76(34.97) + 0.24(36.97) = 35.45$$
$$^{35}Cl \qquad\qquad ^{37}Cl \qquad\qquad \text{natural mixture}$$

Now, let's perform that calculation in the opposite direction. Beginning with the known atomic weight of natural chlorine, determine the abundance of the two isotopes:

$$x = \text{fraction } {}^{37}\text{Cl}$$
$$1 - x = \text{fraction } {}^{35}\text{Cl}$$

Instead of using the integers 37 and 35 as atomic masses, take the more precise atomic masses of the isotopes from the previous chart:

$$36.97(x) + 34.97(1 - x) = 35.45$$
$$36.97x + 34.97 - 34.97x = 35.45$$
$$2.00x = 0.48$$
$$x = 0.24$$
$$1 - x = 1 - 0.24 = 0.76$$

The calculation reveals that natural chlorine is 24% ${}^{37}\text{Cl}$ and 76% ${}^{35}\text{Cl}$.

The most carefully studied element is the simplest, hydrogen, which has a natural atomic mass (1.0080) slightly greater than that of a single proton (1.0078). This mass excess is only 0.0002 atomic mass units, but the investigation of this excess revealed the 3 isotopes of that element.

ISOTOPES OF HYDROGEN

Isotope	Protons	Neutrons	Atomic mass	Natural abundance
${}^{1}\text{H}$	1	0	1.0078	99.985%
${}^{2}\text{H}$	1	1	2.0141	0.015%
${}^{3}\text{H}$	1	2	3.0161	rare

${}^{2}\text{H}$ is often called deuterium, and ${}^{3}\text{H}$ is referred to as tritium. The atomic weight of natural hydrogen (1.0080) exceeds that of ${}^{1}\text{H}$ because of the admixture of deuterium:

$$0.99985\,(1.0078) + 0.00015\,(2.0141) = 1.0080$$

^1H	^2H	natural mixture

Problem 6. In the small chart below, which nuclei are isotopes of one chemical element? Can you give the element's name? Which nuclei have nearly the same mass?

Nucleus	Protons	Neutrons
A	11	13
B	12	12
C	12	13

Problem 7. Listed below are the atomic weights for natural silver and its two isotopes. Use that data to calculate the percentage of silver-109 in the natural mixture.

^{107}Ag	106.905 atomic mass units
^{109}Ag	108.905
natural Ag	107.868

Radioactivity

If you look at the periodic table (page 10), you will notice that all elements after bismuth, atomic number 83, have their atomic weight denoted by an integer within parentheses. Such large nuclei are unstable and undergo spontaneous disintegration by the emission of particles, called **radioactivity**. The atomic weight shown on the periodic table is the mass number of one of the most common isotopes of each radioactive element.

For chemical purposes, the most important types of radiation are **alpha** and **beta particles**. An alpha particle is a $^4_2$He nucleus with 2 protons and 2 neutrons. This nuclear notation uses a subscript to the lower left to record the number of protons, while the superscript to the upper left is the total number of nucleons. The number of protons identifies the chemical element, while the nucleon total is the mass number for the particle or isotope.

If an unstable nucleus emits an alpha particle, its atomic number decreases by 2 and its atomic mass decreases by 4. The decay of thorium-232 provides an example:

$$^{232}_{90}\text{Th} \longrightarrow {}^{228}_{88}\text{Ra} + {}^4_2\text{He}$$

$$\text{thorium-232} \qquad \text{radium-228} \qquad \text{alpha} \\ \text{particle}$$

Notice that radioactive decay actually changes one chemical element to another element, a process referred to as *transmutation*.

The other chemically important mode of radioactive transmutation is provided by negative beta particles, which are electrons emitted from within atomic nuclei, not from the surrounding electronic orbitals. The beta particle arises from the decay of a neutron to a proton:

$$\text{n}^0 \longrightarrow \text{p}^+ + \text{e}^-$$

$$\text{neutron} \qquad \text{proton} \quad \text{electron}$$

The creation of the proton causes the atomic number to increase by one. An example of beta transmutation is the decay of lead-212:

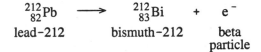

$$^{212}_{82}\text{Pb} \longrightarrow {}^{212}_{83}\text{Bi} + \text{e}^-$$

$$\text{lead-212} \qquad \text{bismuth-212} \qquad \text{beta} \\ \text{particle}$$

The atomic number increased to 83 due to the new proton, but the atomic mass stayed constant because both metal nuclei have 212 total protons plus neutrons.

The most familiar radioactive element is uranium, which has two naturally occurring isotopes of masses 235 and 238 that decay very slowly. Let's review the first few steps in the decay of uranium-238, which changes to lead-206 after the emission of 8 alpha and 6 beta particles. The earliest stages of the decay scheme involve only three elements:

90	91	92
Th	Pa	U
Thorium	Protactinium	Uranium
(232)	(231)	(238)

■ Figure 6 ■

You should carefully examine the changes in both atomic number (protons) and atomic mass (protons plus neutrons):

$$\ce{^{238}_{92}U} \xrightarrow{\alpha} \ce{^{234}_{90}Th} \xrightarrow{\beta} \ce{^{234}_{91}Pa} \xrightarrow{\beta} \ce{^{234}_{92}U} \xrightarrow{\alpha} \ce{^{230}_{90}Th}$$

In this brief sequence, there are 2 different isotopes of uranium and 2 of thorium.

Problem 8. The next step in the uranium-238 decay scheme is the emission of an alpha particle from thorium-230. Describe the atomic mass, atomic number, and element name for the resulting nucleus.

Ions

Of the three major subatomic particles, the negatively charged **electron** was discovered first by the English physicist Joseph Thomson in 1897.

As the structure of atoms was probed, it was realized that these low-mass particles occurred in a large "cloud" around the tiny nucleus, which contained almost all the mass of the atom. A neutral atom has precisely equal numbers of protons (+) and electrons (−). Atoms with a charge imbalance are called **ions**. A positive ion will have lost one or more electrons, whereas a negative ion has gained one or more electrons. Here are three different charge states for copper:

COPPER AND ITS IONS

Description	Charge	Protons	Electrons
Neutral atom	Cu^0	29	29
Cuprous ion	Cu^{1+}	29	28
Cupric ion	Cu^{2+}	29	27

The chemical behavior of the various elements is influenced more by the atomic charge than any other intrinsic property. In the periodic table, elements in a column form analogous compounds because they have the same charge on their ions.

The halogens all tend to gain one electron, giving their ions a characteristic charge of −1. The alkali metals, on the other side of the periodic table, all readily lose one electron, so their ions possess a charge of +1. Charges would be balanced in a compound with the number of alkali metal ions equal to the number of halogen ions; NaCl is such a compound. By contrast, consider the **alkaline earths** in the second column on the left side of the table. These elements are metals that form oxides that have an earthy texture and yield alkaline solutions. The alkaline earths have ions with a charge of +2. For one of them to form a compound with a halogen, there would have to be twice as many halogen ions to balance the +2 charge on the metal ions. Consequently, the correct chemical formula for strontium fluoride is SrF_2.

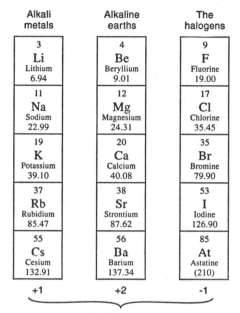

Alkali metals	Alkaline earths	The halogens
3 **Li** Lithium 6.94	4 **Be** Beryllium 9.01	9 **F** Fluorine 19.00
11 **Na** Sodium 22.99	12 **Mg** Magnesium 24.31	17 **Cl** Chlorine 35.45
19 **K** Potassium 39.10	20 **Ca** Calcium 40.08	35 **Br** Bromine 79.90
37 **Rb** Rubidium 85.47	38 **Sr** Strontium 87.62	53 **I** Iodine 126.90
55 **Cs** Cesium 132.91	56 **Ba** Barium 137.34	85 **At** Astatine (210)
+1	+2	-1

Ionic charge

■ Figure 7 ■

Orbitals

Quantum theory assigns the electrons surrounding the nucleus to **orbitals,** which are mathematical devices that should not be confused with the elliptical orbits of the solar system. Each orbital has a characteristic energy and a three-dimensional shape. An atom in the lowest energy configuration is said to be in its **ground state.** For this most stable state, the electrons fill the various orbitals from the one of lowest energy upward. Each orbital may be assigned a maximum of 2 electrons.

The orbitals are completely described by specifying three quantum numbers, but only two of those values are used in this book. The principal quantum number (symbolized n) is a digit 1 or greater that identifies the electron **shell** of the orbital, where the lower digits denote shells of lower energy that are closer to the atomic nucleus. The second quantum number (symbolized l) is a whole number from 0 up to $n - 1$ that defines the shape of the orbital and the maximum number of electrons that can occupy that energy level. For historical reasons, the different shapes of orbitals are represented by letters.

TYPES OF ORBITALS

Second quantum number	Letter denoting orbitals	Number of orbitals	Maximum number electrons
0	s	1	2
1	p	3	6
2	d	5	10
3	f	7	14

Because each orbital holds at most 2 electrons, the maximum number of electrons is twice the number of orbitals with a particular second quantum number. In the preceding chart, *you must know the letters in the second column and the electron capacity in the last column.*

A set of electron orbitals with the same values n and l is called a **subshell** and is represented by notation like $2p^5$.

SUBSHELL NOTATION

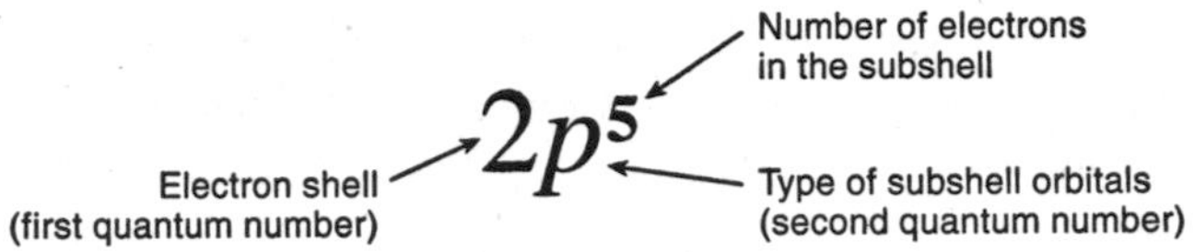

$$2p^5$$

Number of electrons in the subshell

Electron shell (first quantum number)

Type of subshell orbitals (second quantum number)

■ Figure 8 ■

Because the second quantum number must be less than the principal quantum number, only a few subshells are needed to describe the chemical elements in their ground states. The next chart lists all the subshells of chemical importance.

ELECTRON SUBSHELLS

First quantum number	Second quantum number	Notation for subshells
1	0	$1s$
2	0,1	$2s,2p$
3	0,1,2	$3s,3p,3d$
4	0,1,2,3	$4s,4p,4d,4f$
5	0,1,2,3*	$5s,5p,5d,5f$
6	0,1,2*	$6s,6p,6d$
7	0*	$7s$

* Higher orbitals are not occupied in ground state.

Valence Electrons

The electronic configuration of an atom is given by listing its subshells with the number of electrons in each subshell, as in the third column of the chart below. Study the third column of complete electronic configurations carefully so that you will understand how electrons are added to the subshell of lowest energy until it reaches its capacity; then, the subshell of the next energy level begins to be filled.

ELECTRONIC CONFIGURATIONS AND VALENCE

Element name	Atomic number	Electronic configuration	Valence subshell	Common valences
Hydrogen	1	$1s^1$	$1s^1$	+1, −1
Helium	2	$1s^2$	$1s^2$	0
Lithium	3	$1s^2 2s^1$	$2s^1$	+1
Beryllium	4	$1s^2 2s^2$	$2s^2$	+2
Boron	5	$1s^2 2s^2 2p^1$	$2p^1$	+3
Carbon	6	$1s^2 2s^2 2p^2$	$2p^2$	+4, +2, −4
Nitrogen	7	$1s^2 2s^2 2p^3$	$2p^3$	+5, +3, −3
Oxygen	8	$1s^2 2s^2 2p^4$	$2p^4$	−2
Fluorine	9	$1s^2 2s^2 2p^5$	$2p^5$	−1
Neon	10	$1s^2 2s^2 2p^6$	$2p^6$	0
Sodium	11	$1s^2 2s^2 2p^6 3s^1$	$3s^1$	+1
Magnesium	12	$1s^2 2s^2 2p^6 3s^2$	$3s^2$	+2
Aluminum	13	$1s^2 2s^2 2p^6 3s^2 3p^1$	$3p^1$	+3
Silicon	14	$1s^2 2s^2 2p^6 3s^2 3p^2$	$3p^2$	+4
Phosphorous	15	$1s^2 2s^2 2p^6 3s^2 3p^3$	$3p^3$	+5, +3, −3
Sulfur	16	$1s^2 2s^2 2p^6 3s^2 3p^4$	$3p^4$	+6, +4, +2, −2
Chlorine	17	$1s^2 2s^2 2p^6 3s^2 3p^5$	$3p^5$	−1
Argon	18	$1s^2 2s^2 2p^6 3s^2 3p^6$	$3p^6$	0
Potassium	19	$1s^2 2s^2 2p^6 3s^2 3p^6 4s^1$	$4s^1$	+1
Calcium	20	$1s^2 2s^2 2p^6 3s^2 3p^6 4s^2$	$4s^2$	+2

For brevity, many chemists record the electron configuration of an atom by giving only its outermost subshell, like $4s^1$ for potassium or $4s^2$ for calcium. These electrons are most distant from the positive nucleus and, therefore, are most easily transferred between atoms in chemical reactions.

The power of an element to combine with other elements is given by a signed number called the **valence**. For ions, the valence equals the electrical charge. In molecules, the various atoms are assigned charge-like values so that the sum of those valences equals the charge on the molecule. For example, in the H_2O molecule, each H has a valence of +1, and the O is −2.

In the preceding chart, the common valences in the last column are interpreted as the result of either losing the valence electrons (leaving a positive ion) or gaining enough electrons to fill that valence subshell. The next chart compares three ions and a neutral atom.

ELECTRONIC CONFIGURATIONS OF IONS

Chemical element	Valence subshell	Electron transfer	Resulting ion	Ionic configuration
Cl	$3p^5$	gain 1	Cl^-	$1s^2 2s^2 2p^6 3s^2 3p^6$
Ar	$3p^6$	none	Ar^0	$1s^2 2s^2 2p^6 3s^2 3p^6$
K	$4s^1$	lose 1	K^+	$1s^2 2s^2 2p^6 3s^2 3p^6$
Ca	$4s^2$	lose 2	Ca^{2+}	$1s^2 2s^2 2p^6 3s^2 3p^6$

The charges on the chlorine, potassium, and calcium ions result from a strong tendency of valence electrons to adopt the stable configuration of the inert gases, with completely filled electronic shells. Notice that the 3 ions have electronic configurations identical to that of inert argon.

The Periodic Table

The pattern of elements in the periodic table (pages 10-11) reflects the progressive filling of electronic orbitals. The two columns on the left—the alkali metals and alkaline earths—show the addition of 2 electrons into s-type subshells.

FILLING OF s SUBSHELLS

1 H $1s^1$	2 He $1s^2$
3 Li $2s^1$	4 Be $2s^2$
11 Na $3s^1$	12 Mg $3s^2$
19 K $4s^1$	20 Ca $4s^2$

■ Figure 9 ■

The loss of these s-subshell valence electrons explains the common +1 and +2 charges on ions of these metals.

The 6 elements from boron through neon show the insertion of electrons into the lowest-energy subshell of p-type.

FILLING OF p SUBSHELL

5 B $2p^1$	6 C $2p^2$	7 N $2p^3$	8 O $2p^4$	9 F $2p^5$	10 Ne $2p^6$

■ Figure 10 ■

The same type of subshell is used to describe the electron configurations of elements in the underlying rows, like

ANOTHER *p* SUBSHELL

13 Al $3p^1$	14 Si $3p^2$	15 P $3p^3$	16 S $3p^4$	17 Cl $3p^5$	18 Ar $3p^6$

■ Figure 11 ■

The three long rows of metallic elements in the middle of the periodic table, constituting the rectangle from scandium (21) to mercury (80), are the **transition metals**. Each of these three rows reflects the filling of a *d*-type subshell that holds up to 10 electrons. The figure shows the valence subshell of the first series of transition metals. Notice the general increase in the number of electrons occupying the $3d$ subshell.

FILLING OF *d* SUBSHELL

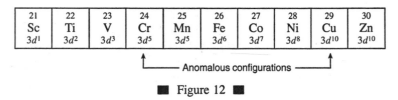

21 Sc $3d^1$	22 Ti $3d^2$	23 V $3d^3$	24 Cr $3d^5$	25 Mn $3d^5$	26 Fe $3d^6$	27 Co $3d^7$	28 Ni $3d^8$	29 Cu $3d^{10}$	30 Zn $3d^{10}$

Anomalous configurations

■ Figure 12 ■

The anomalous electronic configuration of chromium and copper is interpreted as the displacement of 1 electron from an *s* orbital into a *d* orbital; these 2 elements have only 1 electron in the $4s$ subshell because the second electron was promoted into a $3d$ subshell. This example warns you that there are exceptions to the general pattern of electronic configurations of the elements. The complicated electronic structure of the transition metals is a consequence of the similar energy levels of various subshells, like the $4s$ and $3d$ subshells, which leads to multiple valence states for single elements. Vanadium, for example, shows valences of +2, +3, +4, or +5. It also causes ions of the transition metals to produce vivid colors in aqueous solution.

The two rows of **lanthanides** and **actinides** at the bottom of the periodic table belong between elements 57 and 72.

**CORRECT PLACEMENT OF LANTHANIDES
AND ACTINIDES IN THE PERIODIC TABLE**

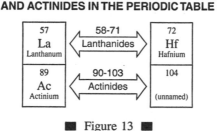

■ Figure 13 ■

Those two long rows of elements are traditionally moved to the base of the chart so that the more important lighter elements may be closer together for clarity. These two rows of metals each reflect the progressive addition of 14 electrons into an f-type subshell. The lanthanides occur in only trace amounts in nature and are often called **rare earths**. All of the actinides have large, unstable nuclei that undergo spontaneous radioactive decay.

The outermost electrons of an atom generally determine the chemical behavior of that element. They determine the atom's size, charge, and ability to exchange electrons with other atoms. If you understand how the periodic table displays the pattern of electron configurations, you will be on your way to mastering chemistry. You should know that rows of the periodic table show filling of various subshells and also that congeners in columns have similar electron subshells that are filled to the same degree.

CHEMICAL BONDING

Covalent Bonds

You will recall from the discussion of electron transfer (page 34) that a stable configuration precisely filled a p-type subshell. Only 6 elements have atoms with their valence p-subshell filled; these are the inert gases in the far right column of the periodic table. Their lack of chemical reactivity is explained by their stable electron configurations.

All other chemical elements need to lose or gain electrons to achieve electronic stability. The following chart shows the stable electron configurations for the elements in the first three rows of the periodic table.

STABLE ELECTRON CONFIGURATIONS

Row in table	Atomic numbers	Electron subshells	Electron capacity	Valence subshells	Valence electrons
1	1–2	$1s^2$	2	$1s^2$	2
2	3–10	$1s^2 2s^2 2p^6$	10	$2s^2 2p^6$	8
3	11–18	$1s^2 2s^2 2p^6 3s^2 3p^6$	18	$3s^2 3p^6$	8

Most atoms achieve a stable number of valence electrons by sharing electrons with other atoms. Let's begin with fluorine, element 9, which has the electron configuration $1s^2 2s^2 2p^5$. The orbitals of the second electron shell are $2s^2 2p^5$, and these 7 valence electrons may be portrayed in a diagram devised by the American chemist Gilbert Lewis (1875-1946). Such a Lewis diagram shows each valence electron as a single dot:

VALENCE ELECTRONS

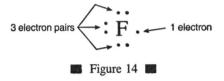

■ Figure 14 ■

Two such fluorine atoms can each fill their valence orbitals with 8 electrons if they approach each other so as to share their single electrons:

SHARING ELECTRONS

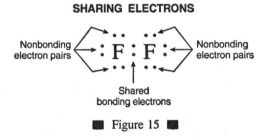

■ Figure 15 ■

Count the electrons in the Lewis diagram to satisfy yourself that there are 14, with each atom contributing 7. The two fluorine atoms form a stable F_2 molecule by sharing electrons; this linkage is called a **covalent bond.**

You can determine the number of valence electrons for the light elements by counting the columns from the left.

COUNTING VALENCE ELECTRONS

11	12	13	14	15	16	17
Na	Mg	Al	Si	P	S	Cl
$3s^1$	$3s^2$	$3s^23p^1$	$3s^23p^2$	$3s^23p^3$	$3s^23p^4$	$3s^23p^5$
1	2	3	4	5	6	7

Valence electrons increase to the right ⟶

■ Figure 16 ■

Phosphorous has 5 valence electrons, and chlorine has 7, so their isolated atoms have these valence configurations:

PHOSPHOROUS AND CHLORINE

■ Figure 17 ■

Phosphorous must combine with 3 chlorines to complete its valence shell:

SIX SHARED ELECTRONS

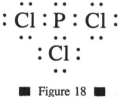

■ Figure 18 ■

Study the figure above carefully. First, see that each atom is now surrounded by a full shell of 8 valence electrons. Of the 26 valence electrons, 6 are shared, and 20 are unshared. For the 6 that are shared to form the covalent bonds, the phosphorous atom contributed 3, and each of the chlorines contributed 1. The resulting PCl_3 molecule is usually drawn as:

THREE COVALENT BONDS

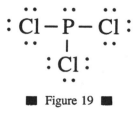

■ Figure 19 ■

In the PCl_3 diagram, each of the three lines represents the shared pair of electrons in a covalent bond. Some textbooks omit the dots representing nonbonding electrons for simplicity.

Because the hydrogen atom has its single $1s$ orbital completed with only 2 electrons, the hydrogen chloride molecule would be drawn as:

HYDROGEN CHLORIDE MOLECULE

$$H : \overset{\cdot\cdot}{\underset{\cdot\cdot}{Cl}} : \quad \text{or} \quad H - \overset{\cdot\cdot}{\underset{\cdot\cdot}{Cl}} :$$

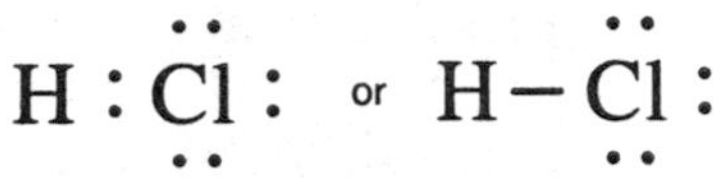

■ Figure 20 ■

The hydrogen and chlorine each donate one electron to the covalent bond. In the molecule, the hydrogen has completed its valence shell with 2 electrons, and the chlorine has a full shell with 8 valence electrons.

In some molecules, bonded atoms share more than 2 electrons, as in ethylene (C_2H_4), where the 2 carbons share 4 electrons.

DOUBLE BOND BETWEEN CARBONS

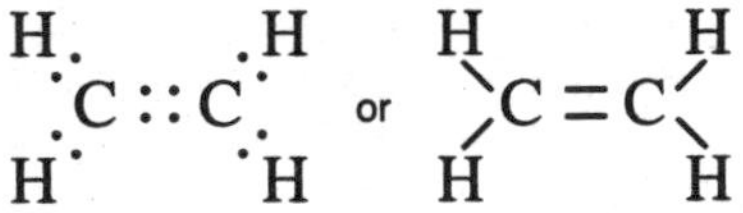

$$\overset{H}{\underset{H}{\cdot}} \, C :: C \, \overset{\cdot H}{\underset{\cdot H}{}} \quad \text{or} \quad \overset{H}{\underset{H}{}} C = C \overset{H}{\underset{H}{}}$$

■ Figure 21 ■

Notice that each carbon achieves 8 electrons by this sharing. Because each shared pair constitutes a single covalent bond, the 2 shared pairs are called a *double bond*. The depiction on the right side of the preceding figure shows this double bond of 4 shared electrons clearly.

There are even *triple bonds* of 6 shared electrons, as in the nitrogen molecule. In N_2, each nitrogen atom contributes 5 valence electrons. Of the 10 electrons shown below, 4 are nonbonding, and 6 comprise the triple bond holding the nitrogen atoms together.

TRIPLE BOND BETWEEN NITROGENS

$$: N :: N : \quad \text{or} \quad : N \equiv N :$$

■ Figure 22 ■

Problem 9. Look at the periodic table and deduce the number of valence electrons for aluminum and oxygen from the positions of the columns for those two elements.

Problem 10. Draw a Lewis diagram representing the electron configuration of the hydrogen sulfide molecule, H_2S.

Ionic Bonds

Although atoms with equal numbers of protons and electrons exhibit no electrical charge, it is common for atoms to attain the stable electronic configuration of the inert gases by either gaining or losing valence electrons. The metallic elements on the left side of the periodic table have electrons in excess of the stable configuration. The next chart shows the electron loss necessary for three light metals to reach a stable electron structure.

IONS OF METALS

Chemical element	Atomic number	Total electrons	Stable number	Electron transfer	Resulting ion
Neon	10	10	10	none	neutral
Sodium	11	11	10	lose 1	Na^+
Magnesium	12	12	10	lose 2	Mg^{2+}
Aluminum	13	13	10	lose 3	Al^{3+}

The positive charge on the resulting metal ion is due to the atom possessing more nuclear protons than orbital electrons. The valence electrons are most distant from the nucleus; thus they are weakly held by the electrostatic attraction of the protons and consequently are easily stripped from atoms of the metals.

By contrast, the nonmetallic elements on the right side of the periodic table have many valence electrons and can most readily attain the stable configuration of the inert gases by gaining electrons. The next chart compares three nonmetals to the inert gas argon.

IONS OF NONMETALS

Chemical element	Atomic number	Total electrons	Stable number	Electron transfer	Resulting ion
Phosphorous	15	15	18	gain 3	P^{3-}
Sulfur	16	16	18	gain 2	S^{2-}
Chlorine	17	17	18	gain 1	Cl^{-}
Argon	18	18	18	none	neutral

Because metallic elements tend to lose electrons and nonmetallic elements tend to gain electrons, a pair of contrasting elements will exchange electrons so that both achieve stable electronic configurations. The resulting ions of opposite charge have a strong force of electrostatic attraction, which is called an **ionic bond**. Note that this bond forms through the complete transfer of electrons from one atom to another, in contrast to the electron sharing of the covalent bond.

The force of attraction between two points of opposite electrical charge is given by Coulomb's Law:

$$\text{force} = \frac{q_+ q_-}{d^2}$$

where q_+ is the positive charge, q_- is the negative charge, and d is the distance between the two charged points. This law of electrostatic attraction can be used to measure the distance between two spherical ions because the charges can be considered to be located at the center of each sphere.

DISTANCE BETWEEN IONIC CHARGES

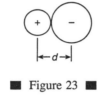

■ Figure 23 ■

Notice that the distance between the centers of the 2 ions is the sum of the ionic radii. The appropriate electrostatic force is then calculated from the equation

$$\text{force} = \frac{q_C q_A}{(r_C + r_A)^2}$$

where q_C is the charge of the positive **cation**, q_A is the charge of the negative **anion**, and the denominator is the sum of their radii.

The strength of ionic bonding, therefore, depends on both the charges and the sizes of the two ions. Higher charges and smaller sizes produce stronger bonds. The next chart shows the approximate radii of selected ions with the electronic configuration of the inert gases. The values for the radii are in Ångstrom units (Å).

IONIC RADII

Cations			Anions	
Li^+ 0.68	Be^{2+} 0.35	B^{3+} 0.23	O^{2-} 1.45	F^- 1.33
Na^+ 0.97	Mg^{2+} 0.66	Al^{3+} 0.51	S^{2-} 1.90	Cl^- 1.81
K^+ 1.33	Ca^{2+} 0.99	Sc^{3+} 0.73	Se^{2-} 2.02	Br^- 1.96
Rb^+ 1.47	Sr^{2+} 1.12	Y^{3+} 0.89	Te^{2-} 2.22	I^- 2.20
Cs^+ 1.67	Ba^{2+} 1.34	La^{3+} 1.06		

Radii in Ångstrom units

The ionic radius increases as you go down any column of uniform charge because the elements of higher atomic number have a greater number of electrons in a series of electronic shells progressively further from the nucleus. The change in ionic size along a row in the chart just above shows the effect of attraction by protons in the nucleus.

The five ions O^{2-} through Al^{3+} studied in the chart below are all **isoelectronic**; that is, they have the same number of electrons in the same orbitals.

VARIATION IN IONIC RADIUS

Ion	O^{2-}	F^-	Ne^0	Na^+	Mg^{2+}	Al^{3+}
Nuclear protons	8	9	10	11	12	13
Orbital electrons	10	10	10	10	10	10
Ionic radius (Å)	1.45	1.33	1.10	0.97	0.66	0.51

For continuity, the neutral Ne atom is also in the chart, with a hypothetical ionic radius. As you proceed to the right in this chart, the greater number of protons attracts the electrons more strongly, producing progressively smaller ions.

Now let's use Coulomb's Law to compare the strengths of the ionic bonds in crystals of magnesium oxide and lithium fluoride. The sizes of the 4 ions are taken from the tabulation of radii of cations and anions on page 46.

$$\text{MgO:} \qquad \frac{(2)(2)}{(0.66 + 1.45)^2} = 0.90$$

$$\text{LiF:} \qquad \frac{(1)(1)}{(0.68 + 1.33)^2} = 0.25$$

Comparing the two relative forces of electrostatic attraction that you calculated, you can conclude that ionic bonding is considerably stronger in magnesium oxide. This affects the physical properties and chemical behavior of the two compounds. For example, the melting point of MgO (2852°C) is much higher than that of LiF (845°C).

The strength of chemical bonding in various substances is commonly measured by the thermal energy (heat) needed to separate the bonded atoms or ions into individual atoms or ions.

Polar Bonds

Many substances contain bonds that are intermediate in character— between pure covalent and pure ionic bonds. Such **polar bonds** occur when one of the elements attracts the shared electrons more strongly than the second element. In potassium iodide, for instance, the shared electrons are so much more attracted by the iodine than the potassium that the sharing is unequal.

UNEQUAL SHARING OF ELECTRONS

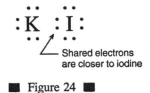

Shared electrons
are closer to iodine

■ Figure 24 ■

Due to the off-center location of the bonding electrons, the 2 atoms
have fractional electrical charges, represented by the Greek letter delta
(δ) in this diagram:

PARTIAL CHARGES OF POLAR BOND

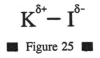

■ Figure 25 ■

Such an off-center covalent bond displays partial ionic character.

The American chemist Linus Pauling developed a scale of **electro-negativity** (ca. 1935) to describe the attraction of the elements for
electrons in a chemical bond. The values in the following chart are
higher for elements that more strongly attract electrons, which would
produce a negative partial charge on that atom.

ELECTRONEGATIVITY

H 2.1	

Li 1.0	Be 1.5	B 2.0	C 2.5	N 3.0	O 3.5	F 4.0
Na 0.9	Mg 1.2	Al 1.5	Si 1.8	P 2.1	S 2.5	Cl 3.0
K 0.8	Ca 1.0	Ga 1.6	Ge 1.8	As 2.0	Se 2.4	Br 2.8
Rb 0.8	Sr 1.0					I 2.4
Cs 0.7	Ba 0.9					

Many elements have been omitted to emphasize the basic pattern of electro-negativity variation.

On the chart just above, you can see that the most electronegative element is fluorine; the nonmetals in the upper right part have a strong tendency to gain electrons. The element of lowest electronegativity is cesium (Cs), in the lower left corner. The relatively weak attraction for electrons by the alkali metals and alkaline earths results in a common loss of electrons by those elements.

Two atoms of the same electronegativity will share electrons equally in a pure covalent bond; therefore, any molecule that contains atoms of only one element, like H_2 or Cl_2, has pure covalent bonding. Two atoms of different electronegativities, however, will have either the distorted electron distribution of a polar bond or the complete electron transfer of an ionic bond. The next chart interprets the bonding between two elements as a function of the difference in their electronegativity.

ELECTRONEGATIVITY AND TYPE OF BOND

Electronegativity difference	Ionic character	Covalent character	Bond type
0.0	0%	100%	covalent
0.5	5	95	covalent
1.0	20	80	covalent
1.5	40	60	polar
2.0	60	40	polar
2.5	75	25	ionic
3.0	90	10	ionic

Now let's use electronegativity to estimate the bond character in hydrogen sulfide, H_2S. The difference in electronegativities is

$$2.5 - 2.1 = 0.4$$
$$S H$$

and you can interpolate this value in the first column of the previous chart to find that such a bond would be about 4% ionic and 96% covalent, which is virtually a pure covalent bond.

Problem 11. Use the chart of electronegativity and the chart of bond types to interpret the bonding in magnesium chloride, $MgCl_2$.

Other Bonds

A polar bond between hydrogen and a nonmetallic element frequently results in a unique secondary bonding between the hydrogen and another negative ion. Such secondary bonding is called a **hydrogen bond** and is much weaker than the primary polar bond. Let's use a water molecule as an example because the electronegativity difference of hydrogen and oxygen is 1.4, indicating that those elements form a polar bond of about 36% ionic character.

POLAR BONDS IN H_2O

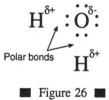

■ Figure 26 ■

The polarity of the hydrogen-oxygen bond leaves the hydrogen atom with its positive nucleus unshielded by any electrons. This positive charge can then electrostatically attract other negative charges, either anions of dissolved salts or the oxygen ($O^{\delta-}$) of other water molecules. In seawater, rich in Na^+ cations and Cl^- anions, the hydrogen attracts the chloride forming hydrogen bonds:

HYDROGEN BONDS

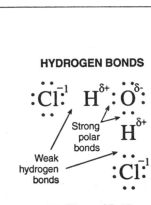

■ Figure 27 ■

This ability of water to form hydrogen bonds explains its unusual property of dissolving many substances.

A final type of bond that you should know about is the **metallic bond** in which electrons move freely among many atoms. The free metals, uncharged metal atoms uncombined with nonmetallic anions, allow this behavior because several electron orbitals usually exist at similar energy levels. An electron is not permanently associated with a single nucleus but may migrate from one to another. This movement of electrons explains why metals are characteristically lustrous, malleable, and highly conductive of heat or electricity.

Structural Formulas

The sixth element in the periodic table, carbon, has the electronic configuration $1s^2 2s^2 2p^2$ and, thus, has 4 valence electrons in the unfilled orbitals of its second electron shell. To fill these orbitals to a stable set of 8 valence electrons, a single carbon atom may share electrons with 1, 2, 3, or even 4 other atoms. No other element forms such strong bonds to as many other atoms as carbon does. Moreover, multiple carbon atoms readily link together with single, double, or triple bonds. These factors make element number 6 unique in the entire periodic table. The number of carbon-based compounds is many times greater than the total of all compounds lacking carbon.

All types of life are based on carbon compounds, and so the study of the chemistry of carbon is called **organic chemistry**. You should realize, however, that "organic compounds" are not necessarily derived from plants and animals. Hundreds of thousands of them have been synthesized (built) in the laboratory from simpler substances.

The following is an illustration of propane, one of the simplest organic compounds:

STRUCTURAL FORMULA FOR PROPANE

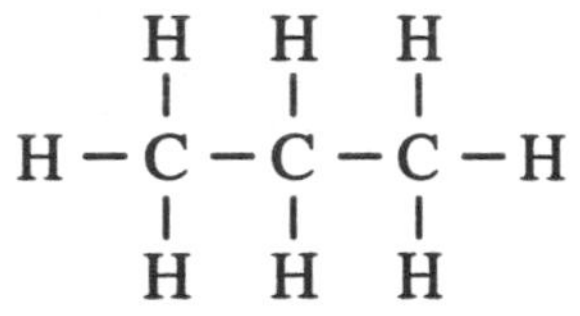

$$
\begin{array}{ccccc}
\text{H} & & \text{H} & & \text{H} \\
| & & | & & | \\
\text{H} - \text{C} & - & \text{C} & - & \text{C} - \text{H} \\
| & & | & & | \\
\text{H} & & \text{H} & & \text{H}
\end{array}
$$

■ Figure 28 ■

This representation is called a **structural formula**; lines depict the bonds between atoms of carbon and hydrogen. Look at the propane structure and observe that the 4 bonds to each carbon complete its valence orbitals with 8 electrons.

In the diagram of propane, the most important feature is the chain of 3 carbons. It is just such carbon-carbon bonding that generates the incredible variety of organic compounds. This linkage of carbon atoms can continue without limit. Just as propane has 3 bonded carbons, you can imagine organic compounds with 4, or 5, or 500 carbons in an extensive chain or network.

The structural formula for propane shows 3 axial carbon atoms and 8 peripheral hydrogen atoms. The composition of propane can be more compactly expressed as C_3H_8; this representation is a **molecular formula**. Such a formula does not directly tell how the various atoms are interbonded.

Let's compare two different compounds that have 4 linked carbon atoms.

ISOMERS OF C_4H_{10}

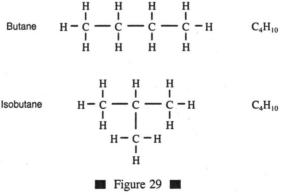

Structural formulas   Molecular formulas

Butane C_4H_{10}

Isobutane C_4H_{10}

■ Figure 29 ■

Although these two compounds have the same molecular formula (and therefore they have identical chemical compositions), their structural formulas reveal a difference in the way that the 4 carbons are assembled. So, *structure is just as essential as composition* to understanding organic chemistry.

The two varieties of C_4H_{10} are called **isomers,** meaning that they have the same composition but differing structures. Structure affects both the physical properties and chemical reactivity of isomers. In the example—C_4H_{10} isomers—they both exist as gases at room temperature, but they can easily be condensed to liquids by cooling or compression. The two liquids have different temperatures at which they boil.

STRUCTURE AND BOILING POINT

Isomer	Boiling point
Butane	$-1°C$
Isobutane	$-12°C$

The boiling behavior is consistent with their structures. The longest carbon chain in butane is 4 atoms, while the longest such chain in isobutane is only 3 atoms. The more compact molecules of isobutane would escape from the liquid more readily, so the more volatile isobutane should have a lower boiling point.

Chemists frequently write condensed structural formulas that omit the plenitude of carbon-hydrogen bonds.

CONDENSED STRUCTURAL FORMULAS

$$CH_3 - CH_2 - CH_2 - CH_3$$

Butane

$$CH_3 - CH - CH_3$$
$$|$$
$$CH_3$$

Isobutane

■ Figure 30 ■

Notice that these condensed structural formulas still display the pattern of carbon-carbon bonding required to distinguish structural isomers.

Hydrocarbons

An infinite variety of compounds can be assembled from only carbon and hydrogen atoms. Such **hydrocarbons** are the simplest organic compounds, but they are also of prime economic importance because they include the constituents of petroleum and natural gas.

Propane, butane, and isobutane are all hydrocarbons with only single covalent bonds between carbon atoms. These hydrocarbons that lack double bonds, triple bonds, or ring structures make up the class called **alkanes.**

THE SIX SIMPLEST ALKANES

Compound name	Molecular formula	Number of isomers
Methane	CH_4	1
Ethane	C_2H_6	1
Propane	C_3H_8	1
Butane	C_4H_{10}	2
Pentane	C_5H_{12}	3
Hexane	C_6H_{14}	5

As the number of carbon atoms increases, so does the number of ways that they can be connected for different isomers. You should realize that isomers are defined by the pattern of bonding between the carbons.

ONLY ONE ISOMER

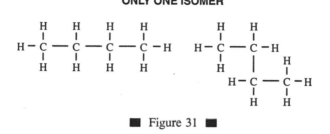

■ Figure 31 ■

The two molecules in the figure are *not* different isomers; they are both butane. Despite the crooked CCCC chain of the molecule on the right, it still has the same condensed structural formula:

$$CH_3 - CH_2 - CH_2 - CH_3$$

■ Figure 32 ■

An **alkene** is a hydrocarbon with at least one double bond between carbons. The simplest alkene is ethylene, C_2H_4.

ALKENE WITH DOUBLE BOND

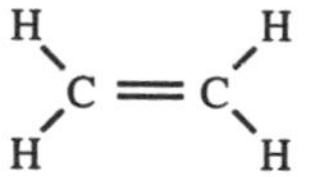

$$\begin{matrix} H & & H \\ \diagdown & & \diagup \\ & C = C \\ \diagup & & \diagdown \\ H & & H \end{matrix} \quad \text{or} \quad CH_2 = CH_2$$

■ Figure 33 ■

As was the case with the alkanes, each carbon atom has precisely 4 bonds to fill its valence orbitals with 8 electrons.

Another simple alkene is propene, C_3H_6.

ALKENE WITH SINGLE AND DOUBLE BONDS

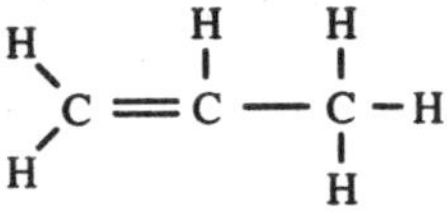

$$\overset{H}{\underset{H}{\diagdown}}C = \overset{\overset{H}{|}}{C} - \overset{\overset{H}{|}}{\underset{\underset{H}{|}}{C}} - H \quad \text{or} \quad CH_2 = CH - CH_3$$

■ Figure 34 ■

Propene demonstrates that alkenes can (and usually do) contain single bonds between some carbons. The existence of any double bond between carbons is the defining character.

A hydrocarbon with a triple bond between carbons is an **alkyne**, and the simplest compound in that class is acetylene, C_2H_2.

ALKENE WITH TRIPLE BOND

$$H - C \equiv C - H \quad \text{or} \quad CH \equiv CH$$

■ Figure 35 ■

Once again, there are exactly four bonds to each carbon. Of course, the triple bond between carbons allows each carbon to bond to only one further atom. In acetylene, the single bond is to hydrogen, but in other alkynes, the single bond is to another carbon. The chart below compares 3 hydrocarbons that contain the same number of carbon atoms.

COMPOUNDS WITH TWO CARBONS

Hydrocarbon class	Compound name	Molecular formula	Carbon bonding
Alkane	ethane	C_2H_6	single
Alkene	ethylene	C_2H_4	double
Alkyne	acetylene	C_2H_2	triple

Look at the third column of the chart and appreciate the diminishing hydrogen content of the compounds as the number of carbon-carbon bonds increases. Organic compounds with multiple carbon-carbon bonds readily react with hydrogen gas.

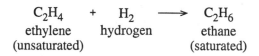

$$C_2H_4 \quad + \quad H_2 \quad \longrightarrow \quad C_2H_6$$
$$\text{ethylene} \quad \text{hydrogen} \quad \text{ethane}$$
$$\text{(unsaturated)} \quad\quad\quad \text{(saturated)}$$

The hydrogenation reaction is possible only for compounds with multiple bonds, and such compounds are said to be *unsaturated*. The addition of the hydrogen to the carbon atoms that were double- or triple-bonded converts the unsaturated compound to a *saturated* compound with only single bonds.

It is possible for long chains of carbons to loop around and form a closed ring structure. If you took the linear isomer of hexane,

HEXANE

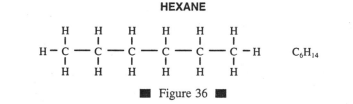

■ Figure 36 ■

and deleted the 2 hydrogens on the ends, the chain could form a hexagonal structure.

CYCLOHEXANE

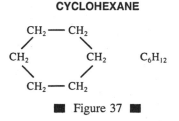

■ Figure 37 ■

Cyclohexane contains only single bonds and is representative of the simplest type of cyclic hydrocarbons.

A ring structure may possess double bonds, as in the following portrayal of the well-known hydrocarbon benzene, which has the composition C_6H_6.

BENZENE

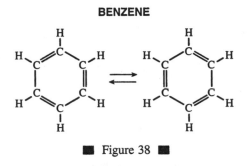

■ Figure 38 ■

The two representations of the benzene ring differ in the location of the 3 double bonds. The arrows between the structures represent hypothetical transitions between the two possible configurations. Only one variety of benzene exists with all 6 carbon-carbon bonds having the same length and strength; so, it seems best to regard the 6 extra electrons of the double bonds as being delocalized over the entire ring structure. Substances with benzene-like rings are called **aromatic** compounds.

Problem 12. Show the 3 isomers of pentane as condensed structural formulas.

Problem 13. Write a balanced molecular reaction for the hydrogenation of acetylene to a saturated alkane. How many liters of hydrogen gas are needed to react completely with 100 liters of acetylene?

Compounds with Additional Elements

The discussion of organic chemistry to this point has described only compounds of carbon and hydrogen. Although all organic compounds contain carbon, and almost all have hydrogen, most of them have other elements as well. The most common extra elements in organic compounds are oxygen, nitrogen, sulfur, and the halogens.

The halogens resemble hydrogen in that they need to form a single covalent bond to achieve electronic stability. Consequently, a halogen atom may replace any hydrogen atom in a hydrocarbon. The next figure shows how fluorine or bromine atoms proxy for hydrogen in methane.

METHANE AND TWO DERIVATIVES

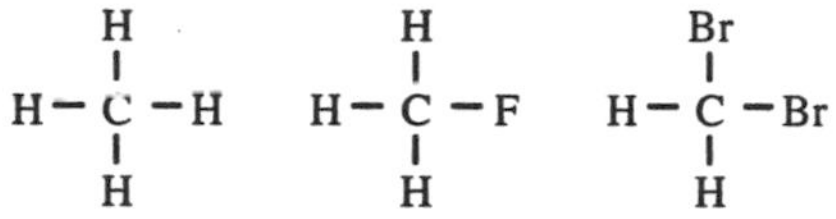

Methane   Methyl fluoride   Methyl dibromide

■ Figure 39 ■

Halogens could replace any or all of the four hydrogens of methane. If the halogen is fluorine, the series of replacement compounds is

$$CH_4 \qquad CH_3F \qquad CH_2F_2 \qquad CHF_3 \qquad CF_4$$

Such halogenated compounds are called organic halides or alkyl halides. The substituted atoms may be fluorine, chlorine, bromine, iodine, or any combination of these elements.

The previously mentioned ethylene molecule is planar. All 6 atoms lie in a single plane because the double bond is rigid.

ETHYLENE

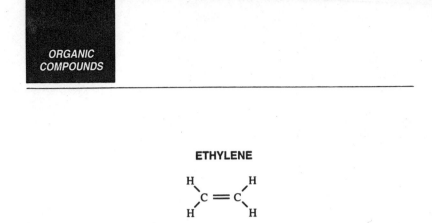

■ Figure 40 ■

The stiff double bond prevents the molecule from being "twisted" around the axis between the carbon atoms. If a reaction substitutes some recognizable object—like a bromine atom—for two of the hydrogens, the resulting compound can exist in either of two different structural configurations.

TWO DIFFERENT MOLECULES

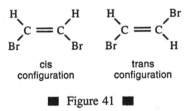

cis
configuration

trans
configuration

■ Figure 41 ■

The configuration with the bromines adjacent is called *cis* (from the Latin for "on this side"), while that with bromines opposite is called *trans* ("on the other side"). The two configurations are different substances with unique chemical and physical properties. They are described as being *geometric isomers*.

The following chart lists some common classes of organic compounds containing oxygen or nitrogen. The main carbon-bearing part of the compound attaches to the bond extending leftward in the second column.

COMMON FUNCTIONAL GROUPS

Compound class	Functional group	Example of group attached to C_2H_5-
Alcohol	$- O - H$	$H - \overset{\overset{\displaystyle H}{\|}}{\underset{\underset{\displaystyle H}{\|}}{C}} - \overset{\overset{\displaystyle H}{\|}}{\underset{\underset{\displaystyle H}{\|}}{C}} - O - H$
Aldehyde	$\overset{\overset{\displaystyle O}{\|\|}}{- C - H}$	$H - \overset{\overset{\displaystyle H}{\|}}{\underset{\underset{\displaystyle H}{\|}}{C}} - \overset{\overset{\displaystyle H}{\|}}{\underset{\underset{\displaystyle H}{\|}}{C}} - \overset{\overset{\displaystyle O}{\|\|}}{C} - H$
Carboxylic acid	$\overset{\overset{\displaystyle O}{\|\|}}{- C - O - H}$	$H - \overset{\overset{\displaystyle H}{\|}}{\underset{\underset{\displaystyle H}{\|}}{C}} - \overset{\overset{\displaystyle H}{\|}}{\underset{\underset{\displaystyle H}{\|}}{C}} - \overset{\overset{\displaystyle O}{\|\|}}{C} - O - H$
Amine	$\underset{\underset{\displaystyle \|}{\|}}{- N -}$	$H - \overset{\overset{\displaystyle H}{\|}}{\underset{\underset{\displaystyle H}{\|}}{C}} - \overset{\overset{\displaystyle H}{\|}}{\underset{\underset{\displaystyle H}{\|}}{C}} - \overset{}{\underset{\underset{\displaystyle H}{\|}}{N}} - H$

■ Figure 42 ■

The examples in the chart all use the ethyl C_2H_5 unit as the carbon chain attached to the functional group, but the immense number of organic compounds arises from the fact that virtually any carbon chain could be attached at that site.

If you compare the carbon-oxygen bonding in Figure 42, you will observe that oxygens may be bonded to carbon by either single or double bonds.

Both alcohols and carboxylic acids have a single hydrogen bonded to an oxygen in the functional group. In aqueous solution, such hydrogens can become detached, producing slightly acidic solutions. More about this property will be discussed later.

The amines contain nitrogen bonded to 1, 2, or 3 carbon chains. These compounds are derivatives of ammonia, hence the name of the class.

AMMONIA

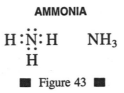

H $:\overset{..}{\underset{..}{N}}:$ H NH_3

     H

■ Figure 43 ■

Consider 3 possible amines created by replacing hydrogen with the CH_3 methyl group.

METHYL DERIVATIVES OF AMMONIA

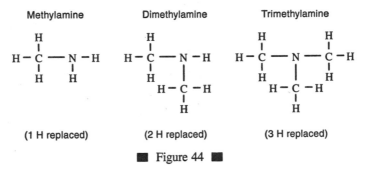

| Methylamine | Dimethylamine | Trimethylamine |

(1 H replaced)   (2 H replaced)   (3 H replaced)

■ Figure 44 ■

Of course, more complex carbon groups could be attached at any of the 3 bonds to nitrogen. Notice that the nitrogen atom is truly the core atom in an amine, in contrast to the functional groups in alcohols, aldehydes, and carboxylic acids, in each of which the functional group must be at the end of the molecule.

Problem 14. The oxidation of methyl alcohol produces a substance that has the composition of CH_2O. Draw the structure of this molecule and classify it on the basis of its functional group.

Solids, Liquids, and Gases

The familiar compound H_2O provides the evidence that substances occur in three different physical classes called **states**. At room temperature, H_2O is a dense fluid called a **liquid**. When that liquid is chilled to $0°C$, it changes to a rigid **solid**. If the liquid is heated to $100°C$, however, it abruptly expands to a tenuous fluid called vapor or **gas**.

Such different states of matter are not unique to H_2O. Almost all substances can exist in two or three of the fundamental states. The chart below defines the states in terms of the shape and volume of substances. Because both liquids and gases flow readily, they are collectively referred to as **fluids**.

DEFINITIONS OF THE STATES OF MATTER

State of matter	Shape of substance	Volume of substance
Solid	definite	definite
Liquid	indefinite	definite
Gas	indefinite	indefinite

These states have different properties because they have distinct structures on the atomic or molecular scale. In a solid, the atoms are strongly bonded to the surrounding atoms so that each is in a fixed position; if the solid structure has a regular pattern that is repeated throughout the solid, it is described as a crystalline structure. The atoms or molecules in a liquid are less strongly bonded than in a solid of the same chemical composition, and consequently, they may shift their positions. The bonds in a liquid are, nevertheless, strong enough so that the atoms stay in contact with surrounding atoms. In a gas, the bonding between individual molecules is so weak that individual molecules may move in all directions, and the absence of cohesive forces allows the vapor to expand throughout any container.

Phase Diagrams

Although the introductory example of H_2O mentioned changes of state caused by varying the temperature, it is known that variation of pressure can also produce such changes. Because in laboratory experiments these two environmental factors—temperature and pressure—can each be varied or held constant, they are referred to as independent variables. The following graph uses those variables as axes that describe the physical conditions at each point in the graph. The vertical axis is the natural logarithm (ln) of the pressure measured in atmospheres.

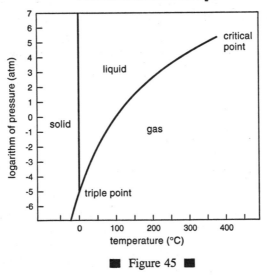

PHASE DIAGRAM FOR H_2O

■ Figure 45 ■

A temperature-pressure graph showing the various states of matter is one type of phase diagram. **Phase** refers to a single homogeneous substance—defined by both its chemical composition and physical state. Different phases have either different compositions or different physical states. In the preceding graph, there are 3 phases with the same composition but different physical states.

Let's begin studying how both temperature and pressure determine the state of H_2O by taking some ice at a temperature of $-10°C$ and pressure of 5 atmospheres, labeled S in Figure 46. If the pressure is held constant but the temperature increased, the substance would heat up along the dashed line marked L, melting to a liquid at point m, about $-0.01°$. Alternatively, if you decreased the pressure on the initial solid S, while holding the temperature constant at $-10°C$, the conditions would change downward along path G, and the ice would vaporize abruptly when the pressure had fallen to the point marked n, about 3×10^{-3} atm. Such a direct change from a solid to a gas is called **sublimation**; notice that there was no intervening liquid state.

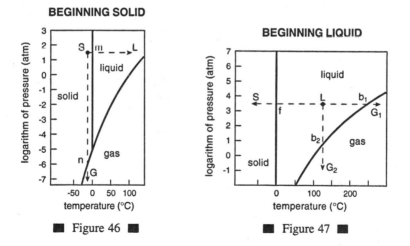

■ Figure 46 ■   ■ Figure 47 ■

In the graph in Figure 47, let's study the possible changes of state of an initial liquid marked L. The liquid is assumed to begin at 120°C and 30 atmospheres. The high pressure allows this liquid to exist at a temperature exceeding the 100°C boiling point at 1 atmosphere. If the pressure is maintained at a constant 30 atm, cooling the liquid L will produce a change to the left along path S, and the liquid will freeze at point f (about -0.01°) to solid ice. A second course with constant pressure is heating L toward G_1, and the liquid would abruptly vaporize at boiling point b_1 (about 235°C). Returning to the initial liquid L, you can imagine holding the temperature constant at 120° while decreasing the pressure toward G_2. When the pressure has fallen to approximately 2 atm, the liquid would boil at point b_2. Boiling has been induced without heating the liquid.

In summary, a change of state can be caused by varying only the temperature, or varying only the pressure, or varying both temperature and pressure. Most random combinations of temperature and pressure fall within the three areas of a phase diagram in which only a single state is stable. The special temperature-pressure combinations plotted as lines in the phase diagram of H_2O are where two states can coexist. For example, both solid ice and liquid water are stable at precisely 0°C and 1 atm.

Look back at the large phase diagram (Figure 45, page 66) and notice the intersection of the three lines at 0.01° and 6×10^{-3} atm. Only at this **triple point** can the solid, liquid, and vapor states of H_2O all coexist. Now, find the point at 374° and 218 atm where the liquid/gas boundary terminates. This **critical point** is the highest temperature and highest pressure at which there is a difference between liquid and gas states. At either a temperature or a pressure over the critical point, only a single fluid state exists, and there is a smooth transition from a dense, liquid-like fluid to a tenuous, gas-like fluid.

Each substance has its own phase diagram to display how temperature and pressure determine its properties. The next graph is the phase diagram for carbon dioxide.

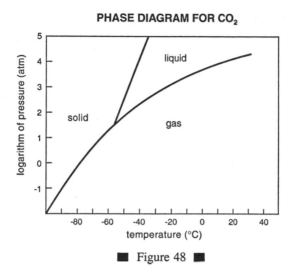

PHASE DIAGRAM FOR CO₂

■ Figure 48 ■

Use the preceding phase diagram to answer the two practice problems.

Problem 15. What is the minimum pressure in atmospheres at which CO_2 can occur as a liquid?

Problem 16. If pressure is held at a uniform 3 atmospheres, at what temperature would solid CO_2 become unstable? What phase begins to appear at that temperature?

Heat Capacities and Transformations

For chemical reactions and phase transformations, the energy required or liberated is commonly studied as **heat**, which is a measure of molecular motion. The principal unit for reporting heat is the **calorie**, which is defined as the energy needed to raise the temperature of

1 gram of water at 14.5°C by a single degree. The term *kilocalorie* refers to 1,000 calories. Another unit of energy is the **joule** (rhymes with "school"), which is equal to 0.239 calories. Conversely, a calorie is 4.184 joules. The translation of calories to joules, or kilocalories to kilojoules, is so common in chemical calculations that you should memorize the conversion factors.

If a substance is heated without a change of state, the amount of heat required to change the temperature of 1 gram by 1°C is called the **specific heat capacity** of the substance. Similarly, the **molar heat capacity** is the amount of heat needed to raise the temperature of 1 mole of a substance by 1°C. The following chart shows the heat capacities of several elements and compounds.

HEAT CAPACITIES

Substance	Calories/degree	
	per gram	per mole
$CaCO_3$	0.205	20.52
H_2O (liquid)	1.000	18.02
H_2O (solid)	0.485	8.74
MgO	0.208	8.38
Pb	0.031	6.32
Fe	0.108	6.01
Al	0.213	5.74

As an example of the use of the heat capacity values, let's calculate the calories required to heat 1 kilogram of aluminum from 10°C to 70°C. Multiply the grams of metal by the 60° increase by the specific heat capacity:

$$1,000 \text{ grams} \times 60 \text{ deg} \times 0.213 \text{ cal/deg-g} = 12,780 \text{ calories}$$

It therefore requires 12.78 kilocalories of energy to heat this particular piece of aluminum. Conversely, if a kilogram of the same metal cooled from 70° to 10°, 12.78 kcal of heat would be released into the environment.

If you examine the two entries for H_2O in the chart of heat capacities, you will realize that there is an abrupt change of energy when one state of matter is transformed into another. A considerable amount of energy is required to transform a low energy state to a higher energy state, like melting a solid to a liquid or vaporizing a liquid to a gas. The same quantity of energy would be released upon the reverse transformation from a high energy state to a lower energy state, like condensing a gas to a liquid or freezing a liquid to a solid. The next chart shows these energy values for H_2O.

HEATS OF TRANSFORMATION FOR H_2O

Change of state	Associated energy	Calories per gram	Calories per mole	Heat
Solid → Liquid	heat of fusion	79.8	1436	absorbed
Liquid → Solid	heat of crystallization	79.8	1436	released
Liquid → Gas	heat of vaporization	539.4	9715	absorbed
Gas → Liquid	heat of condensation	539.4	9715	released

Bear in mind that such transformations of state are isothermal; that is, they take place without any change in temperature of the substance, in this case, H_2O. It takes 79.8 calories to change 1 gram of ice at 0°C to 1 gram of water at 0°C; the 79.8 calories are used to rearrange the molecules from the crystalline order in the solid to the more irregular order in the liquid.

The data in the two preceding charts do permit some complex calculations of energy for changes of both state and temperature. Take a mole of water vapor at 100°C and cool it to ice at 0°. The energy released, which must be removed by the refrigeration process, comes from three distinct changes listed in the chart below.

EXAMPLE OF HEAT CALCULATION

Initial	Final	Calories	Energy source
Vapor @ 100° → Liquid @ 100°		9,715	Heat of condensation
Liquid @ 100° → Liquid @ 0°		1,802	Heat capacity of water
Liquid @ 0° → Solid @ 0°		1,436	Heat of crystallization
		12,953	Total heat released

You should make sure that you understand how each of the values in the third column is obtained. For example, the 1,802 cal is the molar heat capacity of water (18.02 cal/deg) multiplied by the 100 degrees change in temperature.

Notice especially that of the total heat released in this example, only 13.9% comes from lowering the temperature. Most of the heat comes from the two transformations of state—condensation and crystallization. For H_2O, the fact that the heat of condensation is almost 7 times greater than the heat of crystallization may be interpreted as meaning that the molecular description of the liquid state is much more like the solid than the gas.

Problem 17. Use the data in the charts of heat capacity and heat of transformation to calculate the calories required to change 100 grams of ice at −40°C to water at 20°C.

GASES

Boyle's Law

Pressure is the amount of force exerted on one unit of area. The example of an ocean diver should make the concept clearer: the greater the depth the diver reaches, the greater is the pressure due to the weight of the overlying water. Pressure is not unique to liquids but can be transmitted by gases and solids, too. At the surface of the Earth, the weight of the overlying air exerts a pressure equal to that generated by a column of mercury 760 mm high. The two most common units of pressure in chemical studies are atmospheres and millimeters of mercury, where

$$1 \text{ atm} = 760 \text{ mm Hg}$$

The English scientist Robert Boyle performed a series of experiments involving pressure and, in 1662, arrived at a general law—that the volume of a gas varied inversely with pressure.

$$PV = \text{constant}$$

This formulation has become established as **Boyle's Law.** Of course, the relationship is valid only if the temperature is uniform.

As an example of the use of this law, consider an elastic balloon holding 5 liters of air at the normal atmospheric pressure of 760 mm Hg. If an approaching storm caused the pressure to fall to 735 mm Hg, the balloon would expand. The product of the initial pressure and volume is equal to the product of the final pressure and volume because the product is a constant:

$$P_1V_1 = P_2V_2$$

$$(760 \text{ mm})(5 \text{ liters}) = (735 \text{ mm})(x \text{ liters})$$

$$x = \frac{(760)(5)}{735} = 5.17 \text{ liters}$$

It is important that you realize that pressure and volume vary inversely; therefore, an increase in either one necessitates a decrease in the other.

Problem 18. Convert a pressure of 611 mm Hg to atmospheres.

Problem 19. If a gas at 1.13 atm pressure occupies 732 milliliters, what pressure is needed to reduce the volume to 500 milliliters?

Charles' Law

In 1787, the French inventor Jacques Charles, while investigating the inflation of his man-carrying hydrogen balloon, discovered that the volume of a gas varied directly with temperature. This relation can be written

$$\frac{V}{T} = \text{constant}$$

and is called **Charles' Law** (or Gay-Lussac's Law, after the French physicist who first published it). For this law to be valid, the pressure must be held constant and *the temperature must be expressed on the absolute temperature scale.*

Because the volume of a gas decreases with falling temperature, scientists realized that a natural zero-point for temperature could be defined as the volume approaches zero. At a temperature of absolute zero, the volume of an idealized gas would diminish to zero. The absolute temperature scale was devised by the English physicist Kelvin, so temperatures on that scale are called "degrees Kelvin" (°K). The relationship of the Kelvin scale to the common Celsius scale must be memorized by every chemistry student:

$$°K = °C + 273.15$$

Therefore, at normal pressure, water freezes at 273.15°K (0°C) and boils at 373.15°K (100°C). Room temperature is approximately 293°K (20°C). Both temperature scales are used in tables of chemical values, and many simple errors arise from not noticing which scale is presented.

Now let's use Charles' Law to calculate the final volume of a gas, which occupied 400 ml at 20°C and is subsequently heated to 300°C. Begin by converting both temperatures to the absolute scale:

$$T_1 = 20°C = 293.15°K$$

$$T_2 = 300°C = 573.15°K$$

Then, substitute them into the constant ratio of Charles' Law:

$$\frac{V_1}{T_1} = \frac{V_2}{T_2}$$

$$\frac{400 \text{ ml}}{293.15°K} = \frac{x \text{ ml}}{573.15°K}$$

$$x = \frac{(400)(573.15)}{(293.15)} = 782 \text{ ml}$$

In using Charles' Law, it is important to remember that volume and temperature vary directly; therefore, an increase in either requires an increase in the other.

Problem 20. A gas occupying 660 ml at a laboratory temperature of 20°C was refrigerated until it shrank to 125 ml. What is the temperature in degrees Celsius of that chilled gas?

Avogadro's Law

The volume of a gas is determined not only by the pressure and volume but also by its mass. When the mass is given in moles, the mathematical relation is

$$\frac{V}{n} = \text{constant}$$

where n represents the number of moles of the gas. This relationship is known as **Avogadro's Law** because, in 1811, Amedeo Avogadro of Italy proposed that equal volumes of all gases contain the same number of molecules. His theory explained why the volumes of gases in reactions are in ratios of small integers, as in the combustion of hydrogen:

$$2H_2 \quad + \quad O_2 \quad \longrightarrow \quad 2H_2O$$

hydrogen	oxygen	water vapor
(2 volumes)	(1 volume)	(2 volumes)

You should recall from the review of the mole unit that the reaction coefficients (2, 1, and 2 in the example above) can be interpreted as molecules, or as moles, or as volumes. This is because Avogadro's Law holds for all gases that are not severely compressed. The number of molecules in 1 mole of a gas is now known to be 6.02×10^{23}, a value always called **Avogadro's Number**. As an example of the use of this important number, let's calculate the mass in grams of a single oxygen atom:

1 mole O_2 = 2(16.00) = 32.00 grams

1 molecule of $O_2 = \dfrac{32.00}{6.02 \times 10^{23}} = 5.32 \times 10^{-23}$ grams

$$1 \text{ atom of } O = \frac{5.32 \times 10^{-23}}{2} = 2.66 \times 10^{-23} \text{ grams}$$

Problem 21. How many hydrogen atoms are there in 10 grams of methane, CH_4?

Ideal Gas Equation

The relations known as Boyle's Law, Charles' Law, and Avogadro's Law can be combined into an excecdingly useful formula called the **Ideal Gas Equation,**

$$PV = nRT$$

where R denotes the **gas constant:**

$$R = 0.082 \; \frac{\text{liter-atm}}{\text{deg-mole}}$$

The temperature is, as always in gas equations, in degrees Kelvin.

This formula is strictly valid only for ideal gases—those in which the molecules are far enough apart so intermolecular forces can be neglected. At high confining pressures, such forces cause significant departure from the Ideal Gas Equation, and more complicated equations have been devised to treat such cases. The Ideal Gas Equation, however, gives useful results for most gases at pressures less than 100 atmospheres.

The conditions of 0°C temperature and 1 atm pressure may easily be produced in the laboratory; such conditions are called **standard temperature and pressure** (abbreviated **STP**). Because the properties of gases vary with both temperature and pressure, many published values are for gases at STP. Note that room temperature differs from standard temperature.

Let's use the Ideal Gas Equation to calculate the value of 1 mole of an ideal gas at STP:

$$V = \frac{nRT}{P}$$

$$V = \frac{(1 \text{ mole})\left(0.082 \ \dfrac{\text{lit-atm}}{\text{deg-mole}}\right)(273.15 \text{ deg K})}{(1 \text{ atm})}$$

$$V = 22.40 \text{ liters}$$

This is the value stated earlier (page 18); you were asked to memorize that 1 mole of any gas occupies 22.4 liters at STP.

You should be able to use the Ideal Gas Equation to determine any one of the four quantities—pressure, volume, moles, or temperature—if you are given values for the other three.

One important application is to deduce the molecular weight and formula for a gas. Assume you know that the hydrocarbon propylene is, by weight, 85.6% carbon and 14.4% hydrogen. Then the atomic ratios of the compound are

$$C = \frac{85.6}{12.01} = 7.13/7.13 = 1$$

$$H = \frac{14.4}{1.01} = 14.26/7.13 = 2$$

Therefore, the propylene molecule is some integral multiple of CH_2: it could be CH_2 or C_2H_4 or C_3H_6 or a yet larger molecule. Measuring the volume of 10 grams of propylene at STP yields 5.322 liters, which you can use to calculate its molecular weight.

$$\frac{5.322 \text{ liters}}{22.40 \text{ liters/mole}} = 0.2376 \text{ moles}$$

$$\frac{10 \text{ grams}}{0.2376 \text{ moles}} = 42.09 \text{ grams/mole (molecular weight)}$$

Because the atomic weights of 1 CH_2 unit add to 14.03, the molecule must contain 3 such units. Consequently, the molecular formula for propylene is C_3H_6.

Problem 22. What would be the volume occupied by 1 kilogram of carbon monoxide at 700°C and 0.1 atm?

Problem 23. The ozone molecule contains only oxygen atoms. Determine the molecular formula of ozone given that 2.3 grams occupies 1,073 milliliters at standard temperature and pressure.

SOLUTIONS

Concentration Units

A **solution** exists when an abundant host substance contains one or more scarcer substances so evenly dispersed through the host that it appears to be a single, uniform substance. The host substance is the **solvent** and a dispersed substance is a **solute**. Although the most familiar solvents are liquids, like water or ethyl alcohol, the general concept of a solution includes solvents that are gases or even solids.

Essential to the definition of a solution is the capacity of the solvent to contain a variable amount of the solute. Thus, a solution is non-stoichiometric because the ingredients do not have fixed proportions but can occur over a range of values.

Seawater is an example of a liquid solution with water as solvent because the dissolved sodium chloride, calcium carbonate, magnesium bromide, and other solutes are of varying concentrations. Carbonated soda-water is another liquid solution, but in this case, the solute is a gas—carbon dioxide.

Air could be considered to be a gaseous solution with the abundant nitrogen the solvent and scarcer oxygen the solute.

An example of a solid solution is electrum, the alloy of gold and silver. The ratio of the two metals is not fixed but can range from nearly pure gold to nearly pure silver. The dominant metal is deemed to be the solvent in which the minor metal is dissolved.

The abundance of a solute is its **concentration**, and this characteristic can be reported with an intimidating variety of terms that you must master because the accurate description of solutions is central to chemical theory and laboratory practice.

One way to measure the concentration would be by measuring the relative weights of the constituents, usually expressed as *weight percents*. Take, for an example, an electrum ingot that was formed by melting 62 grams of gold and 800 grams of silver, and then letting the material cool and solidify. The following is the composition of the ingot in weight percents:

$$\text{gold:} \quad \frac{62}{862} = 0.072 = 7.2\% \text{ (solute)}$$

$$\text{silver:} \quad \frac{800}{862} = 0.928 = 92.8\% \text{ (solvent)}$$

You know, however, that gold and silver have different atomic weights and the preceding percents do not represent relative numbers of atoms. You could calculate the number of gold and silver moles in the electrum by dividing the weights of the metals by their atomic masses. Those moles allow you to obtain *mole fractions* of the two metals.

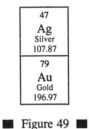

47
Ag
Silver
107.87

79
Au
Gold
196.97

■ Figure 49 ■

$$\text{moles of silver} = \frac{800}{107.87} = 7.416$$

$$\text{moles of gold} = \frac{62}{196.97} = 0.315$$

$$\text{mole fraction silver} = \frac{7.416}{7.731} = 0.959$$

$$\text{mole fraction gold} = \frac{0.315}{7.731} = 0.041$$

Although units of weight percent and mole fraction can be applied to all types of solutions, the familiar solution with water in the role of solvent is most often described by molarity or molality. Such a solution is called an **aqueous** solution.

The **molarity** is the number of moles (or gram formula weights) of solute in 1 liter of solution. This unit is the most convenient one for laboratory work because scaled chemical glassware allows the worker to measure volumes quickly. A solution of calcium chloride that is 0.5 molar (abbreviated 0.5 M with an uppercase "M") contains one-half mole of $CaCl_2$ (55.49 grams) in enough water to make the total volume 1 liter.

The other common unit for liquid solutions is **molality**, the number of moles (or gram formula weights) of solute in 1 kilogram of solvent. Molality contrasts with molarity by reporting the amount of solute relative to the mass of the solvent, not the volume of the solution. A 2 molal solution of hydrogen fluoride, abbreviated 2 m (with a lower-case "m" for distinction from molarity), contains 2 moles of HF (40.02 grams) dissolved in 1,000 grams of H_2O. Molality is the preferred unit for certain types of calculations, although it is less useful in laboratory work. Its merit for calculations arises from the fixed quantity of solvent. In the HF example, the water equals 55.49 moles. Look back at the definition of molarity to satisfy yourself that the solvent is not defined as a fixed amount.

Problem 24. 80 grams of a simple sugar is added to 750 g of water. The sugar is glucose, with the composition $C_6H_{12}O_6$. What is the molality of glucose in the solution?

Problem 25. A solution is prepared by mixing 100 g of methyl alcohol (CH_3OH) and 100 g of water. What is the mole fraction of alcohol in the solution?

Solubility

Although some solutions, like one consisting of water and ethyl alcohol, can have any intermediate composition between the pure components, most solutions have an upper limit to the concentration of the solute. That limit is called the **solubility** of the substance. For example, in a liter of water the maximum amount of $CaSO_4$ that can dissolve is 0.667 grams, which is 0.0049 moles of that solute. Therefore, the solubility of calcium sulfate may be reported either as 0.667 grams per liter or as 0.0049 M.

A solution containing less solute than could dissolve is said to be dilute, and a solution containing as much solute as the solubility limit is described as **saturated**. Adding more of the solute to a saturated solution usually induces some of the solute to separate from the solution; if the separation is by means of the formation of crystals of the solute, those crystals are said to be precipitating from the solution. In some cases, a solution may contain more solute than the solubility limit. But, such a supersaturated state is unstable, and when precipitation begins, it will rapidly lower the concentration of the solute to the saturated level.

The following chart is a useful summary of the relative solubilities of common chemical compounds. They are classified by their anions and listed from the most soluble at the top to the least soluble at the

SOLUBILITIES OF COMPOUNDS

Class	Anion	Description of solubility
Nitrates	NO_3^-	all are highly soluble
Chlorates	ClO_3^-	all are highly soluble
Chlorides	Cl^-	highly soluble except those of silver and mercury
Bromides	Br^-	highly soluble except those of silver and mercury
Iodides	I^-	highly soluble except those of silver, mercury, and lead
Sulfates	SO_4^{2-}	highly soluble except those of strontium, barium, and lead
Sulfides	S_2^{2-}	insoluble except alkali metals, alkaline earths, and ammonium
Sulfites	SO_3^{2-}	insoluble except those of alkali metals and ammonium
Hydroxides	OH^-	insoluble except those of alkali metals and ammonium
Carbonates	CO_3^{2-}	insoluble except those of alkali metals and ammonium
Phosphates	PO_4^{3-}	insoluble except those of alkali metals and ammonium

base. It would also be helpful to remember that all compounds with the cation being an alkali metal or ammonium are highly soluble.

It is important to realize that temperature markedly affects the solubility of most substances. For almost all salts, an increase in temperature leads to an increase in the amount of the salt that will dissolve. The next figure shows the solubilities of potassium chloride (KCl) and potassium nitrate (KNO_3) as a function of temperature.

TEMPERATURE DEPENDENCE OF SOLUBILITY

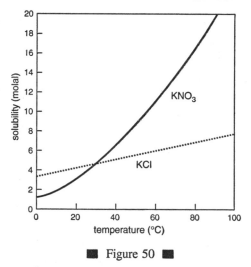

■ Figure 50 ■

Notice that at 20° the KCl is more soluble, but at 40° the KNO_3 has greater solubility. Although the solubilities of both salts increase with temperature, a given temperature rise enhances the solubility of the nitrate much more than the chloride.

When a compound containing ionic bonds is placed in water, the polar water molecules separate some or all of the substance into its cations and anions. The separation is referred to as ionic **dissociation.**

The concentrations of the ions may not equal the concentration of the solution. As an example, take 14.2 grams of sodium sulfate (Na_2SO_4) and add it to enough water to make 1 liter of solution. The sodium sulfate is highly soluble and dissolves completely, so the solution is 0.1 molar in Na_2SO_4.

$$\frac{14.2 \text{ g } Na_2SO_4}{142 \text{ g/mole}} = 0.1 \text{ M}$$

The salt, however, dissociates completely into ions:

$$Na_2SO_4(s) \longrightarrow 2Na^+(aq) + SO_4^{2-}(aq)$$

In the preceding expression, the (s) denotes a solid and (aq) denotes an aqueous ion. In any reaction, the coefficients are proportional to the number of moles. So each mole of Na_2SO_4 yields two moles of Na^+ and one mole of SO_4^{2-}. The 0.1 M solution of Na_2SO_4 is, consequently, 0.2 M in Na^+ and 0.1 M in SO_4^{2-}.

For ionizing substances that are only slightly soluble, the concentrations of the ions multiply to a constant called the **solubility product** in a saturated solution. For a hypothetical compound CA, where the single cation is denoted by C and the anion by A, the solubility equation is

$$[C][A] = K_{sp}$$

where the molar concentrations of the 2 ions are labeled with square brackets and the constant K_{sp} is the solubility product.

Many binary compounds (those with only 2 chemical elements) contain more than 1 cation or anion. The general binary compound could be written C_xA_y where the subscripts mean the compound has x cations and y anions. In this case, the solubility equation is

$$[C]^x[A]^y = K_{sp}$$

Now let's do a solubility calculation using silver carbonate (Ag_2CO_3) as the solute. Dissocation of the salt yields 3 aqueous ions:

$$Ag_2CO_3(s) \longrightarrow 2Ag^+(aq) + CO_3^{2-}(aq)$$

Reference tables state that K_{sp} for Ag_2CO_3 is 8.5×10^{-12}. The solubility equation involves the square of $[Ag^+]$ because each formula unit yields 2 ions of Ag^+.

$$[Ag^+]^2 \ [CO_3^{2-}] = 8.5 \times 10^{-12}$$

Because the molarity of CO_3^{2-} is the same as the overall molarity of Ag_2CO_3 in the solution, call the carbonate concentration x and the silver ion concentration will be $2x$.

$$(2x)^2(x) = 8.5 \times 10^{-2}$$
$$4x^3 = 8.5 \times 10^{-12}$$
$$x^3 = 2.13 \times 10^{-12}$$
$$x = 1.29 \times 10^{-4}$$

The solution, then, is 0.000129 M Ag_2CO_3, which is identical to the value found for the CO_3^{2-} concentration. Because the gram formula weight of Ag_2CO_3 is 275.75, each liter of solution contains

$$0.000129 \text{ moles} \times 275.75 \ \frac{\text{grams}}{\text{mole}} = 0.0355 \text{ grams}$$

and you can describe the solubility of silver carbonate as 0.0355 grams per liter.

The following chart gives the solubility products for some important compounds that are sparingly soluble. The values are for 25°C, and each of them would vary directly with temperature.

SOLUBILITY PRODUCTS AT 25°C

Compound	K_{sp}	Compound	K_{sp}
AgBr	6.3×10^{-13}	CuBr	2.4×10^{-8}
Ag_2CO_3	8.5×10^{-12}	CuCl	1.1×10^{-6}
$Al(OH)_3$	3.7×10^{-15}	$Cu(OH)_2$	2.2×10^{-20}
BaF_2	1.3×10^{-6}	$FeCO_3$	3.2×10^{-11}
$BaSO_4$	1.1×10^{-10}	$MgCO_3$	4.0×10^{-5}
CaF_2	3.9×10^{-11}	MgF_2	6.6×10^{-9}
$CaSO_4$	1.7×10^{-5}	$Mg(OH)_2$	1.5×10^{-4}

You will need to use the chart above to solve both of these practice problems.

Problem 26. Suppose that you stirred 0.15 grams of cuprous chloride (CuCl) powder into a liter of water. Will the powder entirely dissolve?

Problem 27. Determine the solubility in grams per liter for aluminum hydroxide, $Al(OH)_3$.

Freezing and Boiling Points

For a solution with a liquid as solvent, the temperature at which it freezes to a solid is slightly lower than the freezing point of the pure solvent. This phenomenon is known as **freezing point depression** and is related in a simple manner to the concentration of the solute. The lowering of the freezing point is given by

$$\Delta T_f = K_f m$$

where K_f is a constant that depends on the specific solvent and m is the molality of the solute. The next chart gives data for several common solvents.

MOLAL FREEZING POINT AND BOILING POINT CONSTANTS

Solvent	Formula	Freezing point (°C)	K_f (°C/mol)	Boiling point (°C)	K_b (°C/mol)
Water	H_2O	0.0	1.86	100.0	0.51
Acetic acid	CH_3COOH	17.0	3.90	118.1	3.07
Benzene	C_6H_6	5.5	4.90	80.2	2.53
Chloroform	$CHCl_3$	−63.5	4.68	61.2	3.63
Ethanol	C_2H_5OH	−114.7	1.99	78.4	1.22
Phenol	C_6H_5OH	43.0	7.40	181.0	3.56

Let's use the preceding formula and the constant from the chart to calculate the temperature at which a solution of 50 grams of sucrose $(C_{12}H_{22}O_{11})$ in 400 grams of water would freeze. The molecular weight of sucrose is

$$12(12.01) + 22(1.01) + 11(16.00) = 342.34$$

so, the moles of sucrose are

$$\frac{50 \text{ grams}}{342.34 \text{ g/mole}} = 0.146 \text{ moles}$$

and the concentration of the solution in moles per kilogram of water is

$$\frac{0.146 \text{ moles}}{0.4 \text{ Kg } H_2O} = 0.365 \text{ molal}$$

By taking the freezing point constant for water as 1.86 from the chart and then substituting the values into the equation for freezing point depression, you obtain the change in freezing temperature:

$$\Delta T_f = 1.86 \times 0.365 = 0.68°C$$

Because the freezing point of pure water is 0°C, the sucrose solution freezes at −0.68°C.

A similar property of solutions is **boiling point elevation**. A solution boils at a slightly higher temperature than the pure solvent. The change in the boiling point is calculated from

$$\Delta T_b = K_b m$$

where K_b is the molal boiling point constant and m is the concentration of the solute expressed as molality. The boiling point data for some solvents were provided in the chart on page 89.

Notice that the change in freezing or boiling temperature depends solely on the nature of the solvent, not on the identity of the solute.

One valuable use of these relationships is to determine the molecular weight of various dissolved substances. As an example, let's perform such a calculation to find the molecular weight of the organic compound santonic acid, which dissolves in benzene or chloroform. A solution of 50 grams of santonic acid in 300 grams of benzene boils at 81.91°C. Referring to the preceding chart for the boiling point of pure benzene, the boiling point elevation is

$$81.91° - 80.2° = 1.71° = \Delta T_b$$

Rearranging the boiling point equation to yield molality and substituting the molal boiling point constant from the chart, you can derive the molality of the solution:

$$m = \frac{\Delta T_b}{K_b} = \frac{1.71 \text{ deg}}{2.53 \text{ deg/mol}} = 0.676 \text{ molal}$$

That concentration is the number of moles per kilogram benzene, but the solution used only 300 grams of the solvent. The moles of santonic acid is found by

$$0.3 \ Kg \times 0.676 \ mol/Kg = 0.203 \ moles$$

and the molecular weight is calculated

$$\frac{50 \ grams}{0.203 \ moles} = 246.3 \ grams/mole$$

The boiling point of a solution was used to determine that santonic acid has a molecular weight of approximately 246. You could also have found that value by using the freezing point of the solution.

In the two worked examples, the sucrose and santonic acid existed in solution as molecules, rather than dissociating to ions. The latter case requires the total molality of all ionic species. Let's calculate the total ionic molality of a solution of 50 grams of aluminum bromide ($AlBr_3$) in 700 grams of water. Because the gram formula weight of $AlBr_3$ is

$$26.98 + 3(79.90) = 266.68$$

the amount of $AlBr_3$ in the solution is

$$\frac{50 \ grams}{266.68 \ g/mole} = 0.1875 \ moles$$

The concentration of the solution with respect to $AlBr_3$ formula units is

$$\frac{0.1875 \ moles}{0.7 \ Kg \ solvent} = 0.2678 \ molal$$

Each formula unit of the salt, however, yields one Al^{3+} and three Br^- ions:

$$AlBr_3(s) \longrightarrow Al^{3+}(aq) + 3Br^-(aq)$$

So, the concentrations of the ions are

$$[Al^{3+}] = 0.2678 \text{ molal}$$

$$[Br^-] = 3(0.2678) = 0.8035 \text{ molal}$$

$$[Al^{3+}] + [Br^-] = 1.0713 \text{ molal}$$

The total concentration of ions is 4 times that of the salt. The summary value of 1.0713 m is the effective molality of the ionic solution and would be used to calculate the freezing point, boiling point, or molecular weight.

Problem 28. Calculate the boiling point of a solution of 10 grams of sodium chloride in 200 grams of water.

Problem 29. A solution of 100 grams of brucine in 1 Kg chloroform freezes at −64.69°. What is the molecular weight of brucine?

ACIDS AND BASES

The pH Scale

Even distilled water contains some ions because a small fraction of water molecules dissociates to hydrogen and hydroxyl ions:

$$H_2O \; \rightleftharpoons \; H^+ \; + \; OH^-$$

water molecule	hydrogen ion	hydroxyl ion

The double arrows indicate that the reaction proceeds either way. This condition of reciprocal reaction is called chemical **equilibrium**, and its importance to chemistry cannot be overemphasized. An equilibrium state is a stable, balanced condition, and it can be reproduced by many laboratory researchers. It also can be modeled well by simple mathematical equations.

The representation above of the equilibrium between the water molecule and its ions shows that there are the same number of water molecules forming from union of the 2 ions as are dissociating into ions. The concentrations of the hydrogen and hydroxyl ions obey an equilibrium equation:

$$[H^+]\,[OH^-] = K_w$$

where the concentrations are expressed in molarity and K_w is the ion-product constant for water. At a temperature of 25°C, the value of that constant is

$$K_w = 1.0 \times 10^{-14} \; \frac{\text{mole}^2}{\text{liter}^2}$$

In pure water, the concentrations of H^+ and OH^- must be equal because the dissociation of H_2O yields the same number of each of them. You can calculate the ionic concentrations from the equilibrium equation and the ion-product constant:

$$[H^+][OH^-] = 1 \times 10^{-14} \frac{mole^2}{liter^2}$$

$$[H^+] = [OH^-] = x$$

$$x^2 = 1 \times 10^{-14} \frac{mole^2}{liter^2}$$

$$x = 1 \times 10^{-7} \frac{mole}{liter} = [H^+] = [OH^-]$$

The small value for the ion concentrations means that in pure water only 1 in every 555,000,000 molecules has dissociated.

The substances classified as acids and bases affect the H^+ and OH^- concentrations in solutions, but because those two values must multiply to K_w, increasing one ion must depress the other. An acidic solution has more hydrogen ions than hydroxyl ions, whereas a basic—or alkaline—solution has more hydroxyl ions than hydrogen ions.

ION CONCENTRATIONS
AT 25°C

Solution	$[H^+]$	$[OH^-]$
Acidic	$>10^{-7}$	$<10^{-7}$
Neutral	10^{-7}	10^{-7}
Basic	$<10^{-7}$	$>10^{-7}$

The use of scientific notation to describe ion concentrations is somewhat cumbersome, and chemists have agreed to employ a **pH** scale to state the concentration of hydrogen ions (the capital "H" in pH stands for hydrogen). They define pH to be the negative logarithm to base 10 of the hydrogen ion concentration:

$$pH = -\log_{10} [H^+]$$

Thus, the pH of neutral water at 25°C would be calculated

$$pH = -\log(10^{-7}) = -(-7) = +7$$

and so we say that the pH of pure water is 7.

Let's calculate the pH of an acidic solution prepared by adding enough water to 5 grams of hydrochloric acid to make a solution that has a volume of 1 liter.

$$\frac{5 \text{ grams HCl}}{36.46 \text{ g/mole}} = 0.137 \text{ molar}$$

Assuming the hydrochloric acid is completely dissociated,

$$HCl \longrightarrow H^+ + Cl^-$$

the concentration of the hydrogen ions would be

$$[H^+] = 0.137 \text{ M}$$

and the pH of the solution equals

$$pH = -\log(0.137) = 0.86$$

Notice that this acidic pH is less than 7 because the calculation takes the *negative* logarithm of $[H^+]$. Try this calculation now to make sure that you understand this point.

pH VALUES AT 25°C		
Solution	pH	Litmus
Acidic	<7	red
Neutral	7	grey
Basic	>7	blue

The pH of a solution may be estimated from color indicators that change hue with pH, like litmus or phenolphthalein papers. Where precise values are required, an electrical pH meter is utilized.

Problem 30. Calculate the pH of a solution with $[OH^-] = 2.2 \times 10^{-6}$, and classify the solution as acidic, neutral, or alkaline.

Strong and Weak Acids

Substances that dissociate completely into ions when placed in water are referred to as *strong electrolytes* because the high ionic concentration allows an electric current to pass through the solution. Most compounds with ionic bonds behave in this manner; sodium chloride is an example.

By contrast, substances with only covalent bonds—like the simple sugar glucose—do not dissociate at all and exist in solution as molecules. There also are substances—like sodium carbonate (Na_2CO_3)—that contain both ionic and covalent bonding.

BONDING IN Na$_2$CO$_3$

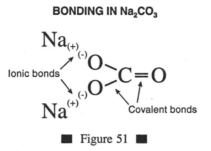

■ Figure 51 ■

The sodium carbonate is a strong electrolyte and dissociates completely to 3 ions when placed in water.

$$Na_2CO_3(s) \longrightarrow 2Na^+(aq) + CO_3^{2-}(aq)$$

| sodium carbonate | sodium cations | carbonate anion |

The carbonate anion is held intact by its internal covalent bonds.

Substances containing polar bonds of intermediate character commonly undergo only partial dissociation when placed in water; such substances are classed as *weak electrolytes*. An example is boric acid:

$$H_3BO_3(s) \longrightarrow H_3BO_3(aq) \rightleftharpoons H^+(aq) + H_2BO_3^-(aq)$$

A solution of that acid is dominated by molecules of H_3BO_3 with relatively scarce H^+ and $H_2BO_3^-$ ions. Make sure you grasp the difference between this case and the previous example of the strong electrolyte Na_2CO_3.

Acids and bases are usefully sorted into *strong* and *weak* classes, depending on their degree of ionization in aqueous solution. Notice that these terms are not defined by the pH of the solution.

The dissociation of any acid may be written as an equilibrium reaction:

$$HA(aq) \rightleftharpoons H^+(aq) + A^-(aq)$$

where A denotes the anion of the particular acid. The concentrations of the three solute species are related by the equilibrium equation

$$\frac{[H^+]\,[A^-]}{[HA]} = K_a$$

where K_a is the **acid ionization constant** (or merely acid constant). Different acids have different K_a values—the higher the value, the greater the degree of dissociation of the acid in solution. Strong acids, therefore, have higher K_a than do weak acids.

The following chart gives acid ionization constants for several familiar acids at 25°C. The values for the strong acids are not well defined. Examine the column labeled "Ions" and see how every acid yields a hydrogen ion and a complementary anion in solution.

SOME COMMON ACIDS

Acid	Formula	Ions	K_a	
Hydrochloric	HCl	H^+ Cl^-	10^7	
Chloric	$HClO_3$	H^+ ClO_3^-	10^3	strong
Sulfuric	H_2SO_4	H^+ HSO_4^-	10^2	acids
Nitric	HNO_3	H^+ NO_3^-	10^1	
Sulfurous	H_2SO_3	$H+$ HSO_3^-	1.5×10^{-2}	
Phosphoric	H_3PO_4	H^+ $H_2PO_4^-$	7.5×10^{-3}	weak
Acetic	CH_3COOH	H^+ CH_3COO^-	1.8×10^{-5}	acids
Carbonic	H_2CO_3	H^+ HCO_3^-	4.3×10^{-7}	
Boric	H_3BO_3	H^+ $H_2BO_3^-$	7.3×10^{-10}	

Let's use the equilibrium equation and data from the preceding chart to calculate the concentrations of solutes in a 1 M solution of carbonic acid. The unknown concentrations of the three species may be written

$$H_2CO_3(aq) \rightleftharpoons H^+(aq) + HCO_3^-(aq)$$
$$1 - x \qquad\qquad x \qquad\quad x$$

where x represents the amount of H_2CO_3 that has dissociated to the pair of ions. Substituting these algebraic values into the equilibrium equation,

$$\frac{[H^+] \, [HCO_3^-]}{[H_2CO_3]} = \frac{x^2}{1 - x} = K_a = 4.3 \times 10^{-7}$$

To solve the quadratic equation by approximation, assume that x is so much less than 1 (carbonic acid is weak and only slightly ionized) that the denominator $1 - x$ may be approximated by 1, yielding the much simpler equation

$$x^2 = 4.3 \times 10^{-7}$$
$$x = 6.56 \times 10^{-4} = [H^+]$$

This H^+ concentration is, as conjectured, much less than the nearly 1 value for the H_2CO_3, so our approximation is valid. A hydrogen ion concentration of 6.56×10^{-4} corresponds to a pH of 3.18.

You will recall from the review of organic chemistry that both alcohols and carboxylic acids have a single hydrogen bonded to an oxygen in the functional group. (See chart on page 63.) To a very small extent, these hydrogens can dissociate in an aqueous solution. Therefore, these two classes of organic compounds are weak acids.

SOME ORGANIC ACIDS

Class	Compound	Formula	K_a
Carboxylic acids	acetic acid	$CH_3 - \overset{\overset{\displaystyle O}{\|\|}}{C} - O - H$	1.6×10^{-5}
	propionic acid	$CH_3CH_2 - \overset{\overset{\displaystyle O}{\|\|}}{C} - O - H$	1.3×10^{-5}
Alcohols	methyl alcohol	$CH_3 - O - H$	3.2×10^{-16}
	ethyl alcohol	$CH_3CH_2 - O - H$	1.0×10^{-16}

■ Figure 52 ■

The four compounds in the preceding chart have been selected because
both acetic acid and methyl alcohol contain the CH_3 group, while
propionic acid and ethyl alcohol both contain the CH_3CH_2 chain.
Although all these compounds behave as weak acids, with dissociation
of the single hydrogen on the right side of the structural formula in the
chart, the much higher K_a values for the first pair in the chart explain
why this class is named carboxylic acids.

Let's summarize the treatment of acids so far. A strong acid is
virtually completely dissociated in aqueous solution, so the H^+
concentration is essentially identical to the concentration of the solution:
for a 0.5 M solution of HCl, $[H^+] = 0.5$ M. But because weak acids

are only slightly dissociated, the concentrations of the ions in such acids must be calculated using the appropriate acid constant.

Problem 31. If an aqueous solution of acetic acid has a pH of 3, how many moles of acetic acid were needed to prepare 1 liter of the solution?

Two Types of Bases

For bases, the concentration of OH^- must exceed that of H^+ in the solution. This imbalance can be created in two different ways.

First, the base can be a hydroxide which merely dissociates to yield hydroxyl ions:

$$MOH \longrightarrow M^+ + OH^-$$

where M represents the cation, usually a metal. The most familiar bases are such hydroxides.

COMMON BASES

Base	Formula	Ions	
Sodium hydroxide	NaOH	Na^+	OH^-
Potassium hydroxide	KOH	K^+	OH^-
Calcium hydroxide	$Ca(OH)_2$	Ca^{2+}	$2OH^-$
Ammonium hydroxide	NH_4OH	NH_4^+	OH^-

The second type of base acts by combining with 1 hydrogen in a water molecule, leaving a hydroxyl ion:

$$B(aq) + H_2O \rightleftharpoons BH^+(aq) + OH^-(aq)$$

base   water   acid   hydroxyl

Alternatively, the base may be an ion with a negative charge:

$$B^-(aq) + H_2O \rightleftharpoons BH(aq) + OH^-(aq)$$

An example of this second type of base that is not a hydroxide would be the ammonia molecule:

$$NH_3(aq) + H_2O \rightleftharpoons NH_4^+(aq) + OH^-(aq)$$

ammonia   ammonium ion   hydroxyl
(base)   (acid)

The ammonia behaved as a base by stripping a proton from a water molecule, resulting in increased OH^- concentration. Notice in the equilibrium reaction that NH_4^+ and NH_3 are a **conjugate** acid-base pair, related by transferring a single proton.

In 1923, the English chemist Thomas Lowry and the Danish chemist Johannes Brönsted defined an acid as a substance that can donate a proton, and a base as a substance that can accept a proton.

Problem 32. The bicarbonate ion HCO_3^- may serve as either a Brönsted-Lowry acid or base. When it acts as an acid, what is its conjugate base? When it behaves as a base, what is its conjugate acid?

Polyprotic Acids

Many acids contain several hydrogen ions. There are 2 in carbonic acid, H_2CO_3, and 3 in phosphoric acid, H_3PO_4. For any such multiple-hydrogen acid, the dissociation of each hydrogen is different. The first hydrogen is most easily removed (the strongest acid), and the last hydrogen is removed with the greatest difficulty (the weakest acid). Because a hydrogen ion is simply a proton, these acids are called **polyprotic** ("many protons") acids. The multiple acid ionization constants for each acid measure the degree of dissociation of the successive hydrogens.

The following chart gives ionization data for four series of polyprotic acids. The integer in parentheses after the name denotes which hydrogen is being ionized, where (1) is the first and most weakly bonded hydrogen.

FOUR SERIES OF POLYPROTIC ACIDS

Acid	Formula	Conjugate base	K_a
Sulfuric (1)	H_2SO_4	HSO_4^-	about 10^{+2}
Sulfuric (2)	HSO_4^-	SO_4^{2-}	1.2×10^{-2}
Sulfurous (1)	H_2SO_3	HSO_3^-	1.6×10^{-2}
Sulfurous (2)	HSO_3^-	SO_3^{2-}	8.3×10^{-8}
Phosphoric (1)	H_3PO_4	$H_2PO_4^-$	7.5×10^{-3}
Phosphoric (2)	$H_2PO_4^-$	HPO_4^{2-}	6.2×10^{-8}
Phosphoric (3)	HPO_4^{2-}	PO_4^{3-}	3.2×10^{-13}
Carbonic (1)	H_2CO_3	HCO_3^-	4.3×10^{-7}
Carbonic (2)	HCO_3^-	CO_3^{2-}	5.2×10^{-11}

Remember that the strongest acids dissociate most readily. Of the 9 acids listed on page 103, the strongest is sulfuric (1), with the highest acid ionization constant, and the weakest is phosphoric (3).

Here are three successive ionizations of phosphoric acid:

First dissociation $\quad H_3PO_4 \rightleftharpoons H^+ + H_2PO_4^-$

Second dissociation $\quad H_2PO_4^- \rightleftharpoons H^+ + HPO_4^{2-}$

Third dissociation $\quad HPO_4^{2-} \rightleftharpoons H^+ + PO_4^{3-}$

Consequently, an aqueous solution of phosphoric acid contains all the following molecules and ions in various concentrations:

$$H_3PO_4 \quad H_2O \quad H^+ \quad OH^- \quad H_2PO_4^- \quad HPO_4^{2-} \quad PO_4^{3-}$$

One of the challenges of solution chemistry is to determine the relative concentrations of the various species. A preliminary deduction may be made from the fact that the solution must be electrically neutral, with total positive charges equal to total negative charges. The sole cation is H^+, which must balance all the anions:

$$[H^+] = [H_2PO_4^-] + 2\,[HPO_4^{2-}] + 3\,[PO_4^{3-}] + [OH^-]$$

Notice that the ion concentrations must be multiplied by the magnitude of the charge. Moreover, because OH^- is only one of several anions in the equation above, it must be true that

$$[H^+] > [OH^-]$$

and, consequently, the solution is acidic.

Look back at the chart of polyprotic acids and satisfy yourself that the K_a values for each of the four acid series change by about 10^{-5} for each successive hydrogen. Therefore, the pH of the solution is determined almost completely by the first dissociation. Let's assume that the pH is set by that first dissociation to calculate the concentrations of all solute molecules and ions.

Now, take a 1 M solution of sulfurous acid for our example. The first dissociation step would be

$$H_2SO_3(aq) \rightleftharpoons H^+ + HSO_3^-$$
$$ 1-x x x$$

where the concentrations have been written algebraically, with x representing $[H^+]$ and $[HSO_3^-]$. You can then substitute these values into the equilibrium equation.

$$K_a(1) = \frac{[H^+][HSO_3^-]}{[H_2SO_3]} = \frac{x^2}{1-x} = 1.6 \times 10^{-2}$$

$$x^2 + 0.016x - 0.016 = 0 \quad \text{(quadratic equation)}$$

$$x = 0.1187 \quad \text{(positive root)}$$

This is the concentration of both H^+ and HSO_3^- in the sulfurous acid solution, and this value will be used in the remaining calculations.

The second dissociation of a small fraction of the HSO_3^- may be represented

$$HSO_3^- \rightleftharpoons H^+ + SO_3^{2-}$$
$$0.1187-y 0.1187+y y$$

where y is SO_3^{2-}. Notice that this second dissociation will increase $[H^+]$ slightly over the value of 0.1187 that was obtained from the first dissociation. The second equilibrium equation is

$$K_a(2) = \frac{[H^+]\,[SO_3^{2-}]}{[HSO_3^-]} = \frac{(0.1187 + y)y}{(0.1187 - y)} = 8.3 \times 10^{-8}$$

Look back at the chart of polyprotic acids and notice that, for sulfurous acid, the ratio of $K_a(2)$ to $K_a(1)$ is approximately 5×10^{-6}. That implies that y is only about 5×10^{-6} of 0.1187, and it can be neglected in the two terms in parentheses, giving a simpler equation:

$$\frac{(0.1187)y}{(0.1187)} = y = 8.3 \times 10^{-8}$$

The concentration of solutes in the solution of sulfurous acid are summarized in this chart.

CONCENTRATIONS IN
SULFUROUS ACID SOLUTION

Solute	Calculation	Molarity
$H_2SO_3(aq)$	$1 - x$	0.8813
HSO_3^-	x	0.1187
SO_3^{2-}	y	8.3×10^{-8}
H^+	x	0.1187
OH^-	$\dfrac{10^{-14}}{x}$	8.4×10^{-14}

From the value of $[H^+]$, you can determine that the pH of the 1 M sulfurous acid is 0.93.

The study of polyprotic acids ought to convince you that the concentrations of the dissolved molecules and ions are interdependent. If you can measure any one of them, all of the others may be derived mathematically.

Problem 33. What is the molar ratio of $H_2PO_4^-$ to PO_4^{3-} in a solution of phosphoric acid with a pH of 4?

Oxidation Numbers

Ions have an electrical charge—negative if they gained electrons and positive if they lost electrons. The existence of ions in crystals or solutions reveals a real exchange of electrons between atoms of different elements. A useful extension of this concept is to assign hypothetical charges, called **oxidation numbers**, to atoms with polar or covalent bonds. The general idea is to assign the shared electrons in each bond to the more electronegative element.

Let's use the water molecule as an example. Here it is in a standard Lewis diagram:

H_2O MOLECULE

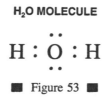

■ Figure 53 ■

Because oxygen is more electronegative than hydrogen, for the purpose of assigning oxidation numbers, it is assumed that all 4 electrons in the 2 covalent bonds are associated with the oxygen atom.

ASSIGNMENT OF ELECTRONS

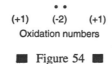

Oxidation numbers

■ Figure 54 ■

The resulting hypothetical electrical charges are the oxidation numbers, which are shown in parentheses to remind you that they are conceptual rather than real. The atoms in H_2O are not ions.

Four rules apply when assigning oxidation numbers to atoms. First, the oxidation number of each atom in a pure element is defined as zero. Therefore, elemental carbon (graphite or diamond) would have an oxidation number of 0, as would an atom in metallic iron, or each of the 2 hydrogen atoms in the H_2 molecule:

$$C^{(0)} \qquad Fe^{(0)} \qquad H_2^{(0)}$$

A single-atom ion is assigned an oxidation number equal to its electrical charge. Examples would be sodium or iron ions, the latter occurring in two oxidation states:

$$Na^{(+1)} \qquad Fe^{(+2)} \qquad Fe^{(+3)}$$

A multiple-atom molecule or ion must have oxidation numbers that sum to the electrical charge of the group of atoms. A neutral molecule has oxidation numbers adding to zero. Therefore, the oxidation numbers of the 1 nitrogen and 3 hydrogens of the neutral NH_3 ammonia molecule sum to 0:

$$N^{(-3)}H_3^{(+1)}$$

while the oxidation numbers of the 1 carbon and 3 oxygens of the charged CO_3^{2-} carbonate ion sum to -2:

$$C^{(+4)}O_3^{(-2)}$$

For each bond between two different elements, the shared electrons are assigned to the element of greater electronegativity, which was nitrogen in the NH_3 example and oxygen in the CO_3^{2-} example. Oxygen usually has an oxidation number of -2, the halogens are commonly -1, hydrogen is almost always $+1$, and the alkali metals are usually $+1$.

Problem 34. What is the oxidation number of nitrogen in magnesium nitride (Mg_3N_2) and in nitric acid (HNO_3)?

Electron Transfer

The concept of oxidation arises from the combination of elemental oxygen with other elements to form oxides, as in this example using aluminum:

$$4Al + 3O_2 \longrightarrow 2Al_2O_3$$

With oxidation numbers inserted as superscripts, this reaction is written

$$4Al^{(0)} + 3O_2^{(0)} \longrightarrow 2Al_2^{(+3)}O_3^{(-2)}$$

which shows that both elements change oxidation numbers. Because the number of electrons associated with each atom determines the oxidation number, an oxidation reaction is defined as one in which electrons are transferred between atoms. In the example, each oxygen atom has gained 2 electrons, and each aluminum has lost 3 electrons.

In an electron transfer reaction, an element undergoing **oxidation** loses electrons, whereas an element gaining electrons is said to undergo **reduction**. In the aluminum-oxygen example, the aluminum was oxidized and the oxygen itself was reduced. Because of electroneutrality on the large scale, *every electron transfer reaction involves simultaneous oxidation and reduction*. These reactions are frequently called **redox** reactions.

A subtlety deserving your close attention is that the oxidizing agent (in the example, oxygen) is reduced, while the reducing agent (in the example, aluminum) is oxidized.

Because chemists have defined oxidation in terms of electron transfer, it is quite unnecessary for redox reactions to have oxygen as the oxidizing agent. Study the next example of metallic zinc reacting with chlorine gas to form zinc chloride:

$$Zn^{(0)} + Cl_2^{(0)} \longrightarrow Zn^{(+2)}Cl_2^{(-1)}$$

The oxidizing agent that gains electrons is the chlorine and the reducing agent that loses electrons is the zinc.

A valuable generalization is that the nonmetals in the upper right of the periodic table are strong oxidizing agents. The metals in their elemental state are strong reducing agents, as is hydrogen gas.

Problem 35. In the following redox reaction, identify the element that is oxidized, the element that is reduced, the oxidizing agent, and the reducing agent.

$$I_2O_5 \quad + \quad 5CO \quad \longrightarrow \quad I_2 \quad + \quad 5CO_2$$

| iodine | carbon | iodine | carbon |
| pentoxide | monoxide | | dioxide |

Balancing Equations

A chemical reaction is said to be balanced when the number of atoms of each element is equal in the reactants and products. Because of the conservation of matter, real reactions are always balanced. You cannot represent the reaction of hydrogen and oxygen as

$$H_2 + O_2 \longrightarrow H_2O \text{ (unbalanced)}$$

because there are more oxygen atoms on the left side. The reaction is correctly written

$$2H_2 + O_2 \longrightarrow 2H_2O \text{ (balanced)}$$

with exactly 4 hydrogen atoms and 2 oxygen atoms on each side. The coefficients of the three substances are chosen so that the reaction is properly balanced.

Although the brief reactions described to this point may be quickly balanced by inspection or trial-and-error, chemistry is rich in complicated reactions that cannot be intuitively balanced. An example would be the use of mercurous hydrogen phosphate in an acidic solution to dissolve gold. Here is the unbalanced reaction without any coefficients:

$$Au(s) + Hg_2HPO_4(s) + H^+(aq) + Cl^-(aq)$$

$$\longrightarrow Hg(l) + AuCl_4^-(aq) + H_2PO_4^-(aq)$$

Take several minutes and try to balance this reaction; the multitude of elements involved makes it a challenging exercise.

Fortunately, most such complicated reactions involve oxidation and reduction, and the oxidation numbers of each element make it much easier to determine the coefficients for a balanced reaction. First, assign oxidation numbers to the elements in each substance. Examine only the elements that change their oxidation number, and insert coefficients so that the overall change is zero. Then modify any coefficients so that the other elements that don't change oxidation number also balance. Finally, check that the electrical charges and the number of atoms of the elements are equal on both sides of the reaction.

As an example, let's use that method to balance the preceding reaction for dissolving gold. Begin by writing oxidation numbers for all the elements in each substance.

$$Au + Hg_2 \, H \, P \, O_4 + H^+ + Cl^- \longrightarrow Hg + Au \, Cl_4^- + H_2 \, P \, O_4^-$$

$$\begin{array}{cccccccc} 0 & +1 & +1 +5 -2 & +1 & -1 & \quad 0 & +3 & -1 & +1 +5 -2 \end{array}$$

oxidation numbers

The oxidation number remains constant for hydrogen, phosphorous, oxygen, and chlorine, and therefore you can initially disregard those elements. Only the gold and mercury show variable oxidation numbers:

loss of 3 electrons

$$Au^{(0)} + Hg^{(+1)} \longrightarrow Hg^{(0)} + Au^{(+3)}$$

gain of 1 electron

The balanced reaction must have equal loss and gain of electrons, so there must be three times as many Hg as Au atoms. Because the mercurous reactant has 2 Hg atoms per formula unit, you would balance the change in oxidation numbers by transferring 6 electrons:

loss of 6 electrons

$$2Au^{(0)} + 3Hg_2^{(+1)} \longrightarrow 6Hg^{(0)} + 2Au^{(+3)}$$

gain of 6 electrons

Inserting the preceding four coefficients into the overall reaction yields

$$2Au + 3Hg_2HPO_4 + H^+ + Cl^- \longrightarrow 6Hg + 2AuCl_4^- + H_2PO_4^-$$

But you are not finished because the numbers of H, P, O, and Cl do not balance. You can, however, balance both P and O by inserting a coefficient of 3 before $H_2PO_4^-$ and balance Cl with a coefficient of 8 before Cl^- on the left:

$$2Au + 3Hg_2HPO_4 + H^+ + 8Cl^- \longrightarrow 6Hg + 2AuCl_4^- + 3H_2PO_4^-$$

The only element that is still unbalanced is H, with 4 on the left side and 6 on the right. (Count them.) A coefficient of 3 before H^+ on the left fixes this imbalance:

$$2Au + 3Hg_2HPO_4 + 3H^+ + 8Cl^- \longrightarrow 6Hg + 2AuCl_4^- + 3H_2PO_4^-$$

This is a balanced redox reaction. You should check that the electrical charges on each side are equal (both 5−) and recount the numbers of atoms for all 6 elements to ensure that they balance.

The preceding technique for balancing reactions is useful because you begin by considering only the few elements involved in electron transfer.

Problem 36. Use oxidation numbers to balance this redox reaction.

$$MnO_4^- + H^+ \longrightarrow MnO_2 + O_2 + H_2O$$

Electric Cells

You have seen that electron transfer occurs in many chemical reactions. This chapter describes how such redox reactions can generate a flow of electricity and, conversely, how electrical currents can induce chemical reactions.

A device that uses a chemical reaction to produce electricity is called an **electric cell**. Because the liquid state allows reactions to occur more readily than either solids or gases, most electrochemical cells are built around a liquid called an **electrolyte**, which conducts electricity. This word has previously been mentioned with regard to ionic dissociation. Pure, distilled water is a very poor conductor of electricity, but a high concentration of dissolved ions leads to high conductivity. That is why acids, bases, and salts which ionize to a high degree are referred to as strong electrolytes, while those which ionize only slightly are referred to as weak electrolytes.

A simple electric cell can be made from two test tubes connected with a third tube (the crossbar of the "H") as shown in the diagram on the next page. The hollow apparatus is filled by simultaneously pouring different solutions into the two test tubes, an aqueous solution of zinc sulfate into the left tube and a copper sulfate solution into the one on the right. Then, a strip of zinc metal is dipped into the $ZnSO_4$ solution, a piece of copper is inserted into the $CuSO_4$ solution, and the two ends of the metal strips are connected by wires to an electrometer. The lateral connecting tube allows ionic migration necessary for a closed electrical circuit. The electrometer will detect the movement of electrons in the wire from the zinc electrode toward the copper electrode; it will also record a potential of 1.10 volts.

AN ELECTRIC CELL

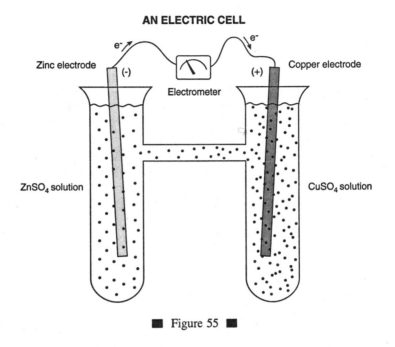

■ Figure 55 ■

The electric current is caused by a pair of redox reactions. At the zinc electrode, the metallic zinc is slowly being dissolved by an oxidation reaction:

$$\text{Zn}(s) \longrightarrow \underset{\text{ion}}{\text{Zn}^{2+}(aq)} + \underset{\text{electrons}}{2e^-}$$
$$\underset{\text{metal}}{}$$

An electrode at which oxidation occurs is called an **anode**; it strongly attracts negative ions in the solution, and such ions are consequently called anions.

Simultaneously, a reduction reaction at the copper **cathode** causes Cu^{2+} cations to be deposited onto the electrode as new copper metal:

$$Cu^{2+}(aq) + \quad 2e^- \quad \longrightarrow \quad Cu(s)$$

$$\text{ion} \qquad \quad \text{electrons} \qquad \text{metal}$$

Because negatively charged electrons are flowing from the anode toward the electrometer, that is the negative electrode. The cathode is the positive electrode.

Adding the redox reactions at the two electrodes gives

$$Zn(s) + Cu^{2+}(aq) \longrightarrow Cu(s) + Zn^{2+}(aq)$$

and that is the overall reaction in the zinc-copper cell.

Electrode Potential

The potential difference, which is measured in volts, depends upon the particular substances constituting the electrodes. For any electric cell, the total potential is the sum of those produced by the reactions at the two electrodes:

$$EMF_{cell} = EMF_{oxidation} + EMF_{reduction}$$

The **EMF** denotes electromotive force, another name for electrical potential.

Chemists have measured the voltages of a great variety of electrodes by connecting each in a cell with a standard hydrogen electrode, which is hydrogen gas at 1 atm bubbling over a platinum wire. This standard electrode is arbitrarily assigned a potential of 0 volts, and measurement of the EMF of the complete cell allows the potential of the other electrode to be determined. The next chart lists some standard potentials for electrodes at which reduction is occurring.

STANDARD ELECTRODE POTENTIALS

Volts	Reduction half-reaction				
2.87	$F_2(g)$	+	$2e^-$	$\longrightarrow$	$2F^-(aq)$
1.36	$Cl_2(g)$	+	$2e^-$	$\longrightarrow$	$2Cl^-(aq)$
1.20	$Pt^{2+}(aq)$	+	$2e^-$	$\longrightarrow$	$Pt(s)$
0.92	$Hg^{2+}(aq)$	+	$2e^-$	$\longrightarrow$	$Hg(l)$
0.80	$Ag^+(aq)$	+	e^-	$\longrightarrow$	$Ag(s)$
0.53	$I_2(s)$	+	$2e^-$	$\longrightarrow$	$2I^-(aq)$
0.34	$Cu^{2+}(aq)$	+	$2e^-$	$\longrightarrow$	$Cu(s)$
0	$2H^+(aq)$	+	$2e^-$	$\longrightarrow$	$H_2(g)$
-0.13	$Pb^{2+}(aq)$	+	$2e^-$	$\longrightarrow$	$Pb(s)$
-0.26	$Ni^{2+}(aq)$	+	$2e^-$	$\longrightarrow$	$Ni(s)$
-0.44	$Fe^{2+}(aq)$	+	$2e^-$	$\longrightarrow$	$Fe(s)$
-0.76	$Zn^{2+}(aq)$	+	$2e^-$	$\longrightarrow$	$Zn(s)$
-1.66	$Al^{3+}(aq)$	+	$3e^-$	$\longrightarrow$	$Al(s)$
-2.71	$Na^+(aq)$	+	e^-	$\longrightarrow$	$Na(s)$
-2.87	$Ca^{2+}(aq)$	+	$2e^-$	$\longrightarrow$	$Ca(s)$
-2.93	$K^+(aq)$	+	e^-	$\longrightarrow$	$K(s)$
-3.04	$Li^+(aq)$	+	e^-	$\longrightarrow$	$Li(s)$

Near the middle of the list, you will see 0 volts arbitrarily assigned to the standard hydrogen electrode; all the other potentials are relative to that hydrogen half-reaction. The voltages are given signs appropriate for a reduction reaction. For oxidation, the sign would be reversed; thus, the oxidation half-reaction,

$$Pt(s) \longrightarrow Pt^{2+}(aq) + 2e^-$$

would have an EMF of −1.20 volts, the opposite to that given in the chart above. Look this up in the chart to be sure you understand.

Let's see how these standard potentials are used to determine the voltage of an electric cell. In the zinc-copper cell described earlier, the two half-reactions must be added to determine the cell EMF.

ZINC-COPPER CELL

Half-reaction	Type	Electrode	Potential
$Zn(s) \longrightarrow Zn^{2+}(aq) + 2e^-$	oxidation	anode	0.76 volts
$Cu^{2+}(aq) + 2e^- \longrightarrow Cu(s)$	reduction	cathode	0.34 volts

The complete zinc-copper cell has a total potential of 1.10 volts (the sum of 0.76 and 0.34). Notice that the sign of the potential of the zinc anode is the reverse of that given in the chart of standard electrode potentials (page 120) because the reaction at the anode is oxidation.

In the chart of standard electrode potentials, reactions are arranged in order of their tendency to occur. Reactions with a positive EMF occur spontaneously, while reactions with a negative EMF require an external electrical potential. The zinc-copper cell had an overall EMF of +1.10 volts, so the solution of zinc and deposition of copper would proceed.

Now let's calculate the total potential of a similar cell with zinc and aluminum electrodes. The next chart shows the two pertinent half-reactions.

ALUMINUM-ZINC CELL

Half-reaction	Type	Electrode	Potential
$2Al(s) \longrightarrow 2Al^{3+}(aq) + 6e^-$	oxidation	anode	1.66 volts
$3Zn^{2+}(aq) + 6e^- \longrightarrow 3Zn(s)$	reduction	cathode	−0.76 volts

Such a cell with zinc and aluminum electrodes would have an overall potential of +0.90 volts, with aluminum being dissolved and zinc metal being deposited out of solution.

If you select any two half-reactions from the chart of standard electrode potentials, the half-reaction higher on the list will proceed in the reduction sense, and the one lower on the list will proceed in the reverse direction as an oxidation. Beware: some references give standard electrode potentials for oxidation half-reactions, so you would have to switch "higher" and "lower" in the rule stated in the preceding sentence.

Problem 37. Some silver mines dump shredded steel cans into ponds containing dissolved silver salts. Write the two redox half-reactions and the overall balanced reaction that explains the deposition of silver from the solution.

Problem 38. Considering only the elements in the chart of standard electrode potentials (on page 120), which pair would make a battery with the greatest voltage? What would the voltage be?

Faraday's Laws

The electric cell with zinc and copper electrodes had an overall potential difference that was positive (+1.10 volts), so the spontaneous chemical reactions produced an electric current. Such a cell is called a Voltaic cell after the Italian physicist Alessandro Volta, who published the first description of one in 1800. In contrast, there are *electrolytic cells* that use an externally generated electrical current to produce a chemical reaction that would not otherwise take place.

An instance of such **electrolysis** is the decomposition of water to elemental hydrogen and oxygen. The pertinent half-reactions are given in the following chart.

ELECTROLYSIS OF WATER

Half-reaction	Type	Electrode	Potential
$4H_2O(l) + 4e^- \longrightarrow 2H_2(g) + 4OH^-(aq)$	reduction	cathode	−0.83 v
$4OH^-(aq) \longrightarrow O_2(g) + 2H_2O(l) + 4e^-$	oxidation	anode	−0.40 v

The overall reaction for the electrolysis of water is given by adding the two half-reactions to obtain

$$2H_2O(l) \longrightarrow 2H_2(g) + O_2(g)$$

with an overall potential of −1.23 volts. With a negative potential, it requires an externally imposed electrical current to decompose water by the reaction shown above. The following figure shows two platinum electrodes in water containing a little salt or acid so that the solution can conduct electricity.

ELECTROLYSIS OF WATER

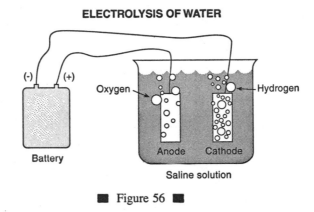

■ Figure 56 ■

The reduction at the cathode yields H_2 gas, and the oxidation at the anode yields O_2 gas. Notice that the figure shows that the volume of hydrogen is twice that of oxygen—look at the bubbles. The molar coefficients in the decomposition reaction imply 2 volumes of H_2 gas for each 1 volume of O_2 gas.

Electrolysis is used to decompose many compounds into their constituent elements. You have seen that with water. Another example is the electrolysis of molten sodium chloride to yield molten sodium metal and chlorine gas:

$$2NaCl(l) \longrightarrow 2Na(l) + Cl_2(g)$$

Chemists throughout the nineteenth century discovered new elements from the electrolytic decomposition of many compounds.

The quantitative laws of electrochemistry were discovered by Michael Faraday of England. His 1834 paper on electrolysis introduced many of the terms that you have seen throughout this book, including *ion, cation, anion, electrode, cathode, anode,* and *electrolyte*. He found that the mass of a substance produced by a redox reaction at an electrode is proportional to the quantity of electrical charge that has passed through the electric cell. For elements with different oxidation numbers, a uniform quantity of electricity produces fewer moles of the element with higher oxidation number.

The basic unit of electrical charge used by chemists is appropriately called a **faraday**, which is defined as the charge on one mole of electrons (6×10^{23} electrons). Incidentally, note that chemists have extended the original definition of the mole as a unit of mass to a corresponding number (Avogadro's Number) of particles. Let's use the electrolysis of molten sodium chloride to see the relationship between faradays of electricity and moles of decomposition products.

The reduction half-reaction is

$$Na^+(l) + e^- \longrightarrow Na(l)$$

so to produce 1 mole of sodium metal requires 1 mole of electrons: 1 faraday of charge must pass through the cell.

The oxidation half-reaction is

$$2Cl^-(l) \longrightarrow Cl_2(g) + 2e^-$$

and to produce 1 mole of chlorine gas, 2 faradays of electric charge must pass through the apparatus. Notice how the number of electrons in redox reactions determines the quantity of electricity needed for the reaction.

These half-reactions sum to the overall reaction in the electrolytic cell:

$$2Na^+(l) + 2Cl^-(l) \longrightarrow 2Na(l) + Cl_2(g)$$

The passage of 2 faradays of charge yields 2 moles of sodium metal and 1 mole of chlorine gas.

One of Michael Faraday's laws was that the mass of substance produced is proportional to the quantity of electricity. To apply this law to the NaCl example, where 1 mole of Cl_2 was produced by 2 faradays, means that to produce 10 moles of Cl_2 would require the passage of 20 faradays through the apparatus.

The other electrolytic law formulated by Faraday states that a given quantity of electricity produces fewer moles of substances with higher oxidation numbers. Compare the reduction of sodium and calcium ions:

$$Na^+ \ + \ e^- \longrightarrow Na \ \text{(1 mole of electrons)}$$

$$Ca^{2+} \ + \ 2e^- \longrightarrow Ca \ \text{(2 moles of electrons)}$$

The electrolytic decomposition of sodium chloride and calcium oxide appear similar:

$$2NaCl \longrightarrow 2Na \ + \ Cl_2$$

$$2CaO \longrightarrow 2Ca \ + \ O_2$$

but the NaCl decomposition requires the transfer of only half as many electrons as does the CaO decomposition. For the NaCl electrolysis, it was previously calculated that the passage of 20 faradays of electric charge would produce 20 moles of sodium metal and 10 moles of chlorine gas. The same amount of electric charge passing through the CaO cell would yield only 10 moles of calcium metal and 5 moles of oxygen gas.

Problem 39. Some aluminum metal is produced by the electrolysis of molten cryolite, Na_3AlF_6. How many faradays of electric charge are needed to produce 1 kilogram of aluminum?

Two Reaction Directions

Earlier in this book, your attention was drawn to the idea of equilibrium and the central role that it plays in chemistry. It was stated that equilibrium is a stable, balanced condition. Let's amplify your understanding of chemical equilibrium through three different examples.

You are aware that the melting of ice can be represented by the reaction

$$H_2O\,(s) \longrightarrow H_2O\,(l)$$

and that the freezing of water can be represented by the reverse reaction

$$H_2O\,(l) \longrightarrow H_2O\,(s)$$

Either of these unilateral reactions is written with the implication that the reactant on the left is completely converted to the product on the right. The situation where both states of H_2O are in equilibrium is shown by the reversible reaction

$$H_2O\,(s) \rightleftharpoons H_2O\,(l)$$

where the two arrows mean that some H_2O molecules are participating in the forward (melting) reaction and other molecules are simultaneously participating in the backward (freezing) reaction. Therefore, equilibrium is the stable situation resulting from two offsetting reactions. At a pressure of 1 atmosphere and a temperature of 0°C, both solid ice and liquid water are stable and may coexist. Notice that this equilibrium condition could have been reached from either side; it could have begun with either pure ice at −10°C or pure water at 20°C.

Whether you warmed such ice or cooled such water, the second phase would appear at 0°C.

The second example will demonstrate another aspect of equilibrium by using the transformations between dinitrogen tetroxide and nitrogen dioxide.

$$N_2O_4(g) \;\rightleftharpoons\; 2NO_2(g)$$

dinitrogen   nitrogen

tetroxide   dioxide

Notice that both gases have the same chemical composition by either atomic or weight percentages of the two elements. N_2O_4 is a colorless gas, while NO_2 is dark reddish-brown. Their relative abundances are a function of temperature because N_2O_4 dominates at room temperature and NO_2 dominates at higher temperatures. At any one temperature, both gases are present in a mixture, and the color of the mixture allows an estimation of the ratio of the two nitrogen oxides. If a glass vessel containing them is colorless or pale, then N_2O_4 exceeds NO_2. Warming that container would cause the color to slowly darken as N_2O_4 is converted to NO_2 until the ratio of the two species is appropriate for the higher temperature.

NITROGEN OXIDES

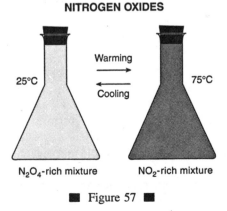

Figure 57

Then the color ceases to change and remains at the new, darker hue. The color change in this gaseous reaction allows you to readily imagine the pair of reactions involved. The experimental fact that warming the gas mixture causes the color to darken shows that temporarily the reaction

$$N_2O_4(g) \longrightarrow 2NO_2(g)$$

dominates over the reverse reaction. The additional consequence that the color stops getting darker shows that a new chemical equilibrium has been reached:

$$N_2O_4(g) \rightleftharpoons 2NO_2(g)$$

in which the forward and reverse reactions occur at the same rate. The rate of creation of NO_2 is precisely equal to the rate of its reaction back to N_2O_4.

The final example demonstrates that equilibrium may involve more than two substances. Sodium iodide will react with sulfuric acid until this chemical equilibrium is established:

$$2NaI(s) \ + \ 2H_2SO_4(l) \rightleftharpoons Na_2SO_4(s) \ + \ 2H_2O(l) \ + \ I_2(g) \ + \ SO_2(g)$$

| sodium iodide | sulfuric acid | sodium sulfate | water | iodine vapor | sulfur dioxide |

At equilibrium, all 6 substances would be present: 2 solid compounds, 2 liquid compounds, 1 gaseous element, and 1 gaseous compound.

Equilibrium Concentrations

In a system that has reached chemical equilibrium, the concentrations of the various substances are quantitatively related. In the example of transformations between the 2 nitrogen oxides, the concentrations obey this equilibrium equation:

$$\frac{[NO_2]^2}{[N_2O_4]} = K$$

where the value K is the **equilibrium constant**.

For any equilibrium reaction, the ratio of concentrations of the substances on the right to the concentrations of those on the left equals a constant appropriate for that specific reaction. Notice that the ratio is always written with the products over the reactants. Each concentration must be raised to the power of its stoichiometric coefficient in the reaction. Because NO_2 has a coefficient of 2 in the reaction, its concentration must be squared in the equation above.

For a generalized reaction written

$$w A + x B \rightleftharpoons y C + z D$$

where the lower-case letters represent numerical coefficients for the balanced reaction, the equilibrium constant would be calculated

$$\frac{[C]^y [D]^z}{[A]^w [B]^x} = K$$

where the brackets denote concentrations of the various substances. Therefore, the reaction of nitrogen and hydrogen to yield ammonia

$$N_2(g) + 3H_2(g) \rightleftharpoons 2NH_3(g)$$

would obey this equilibrium equation:

$$\frac{[NH_3]^2}{[N_2][H_2]^3} = K$$

The dissociation of water vapor to hydrogen and oxygen

$$2H_2O(g) \rightleftharpoons 2H_2(g) + O_2(g)$$

would follow this equation:

$$\frac{[H_2]^2\,[O_2]}{[H_2O]^2} = K$$

Of course, the equilibrium constant K in the latter equation would not have the same value as the K in the equilibrium equation for ammonia. The numerical value of K depends on the particular reaction, the temperature, and the units used to describe concentration. For liquid solutions, the concentrations are usually expressed as molarity. For a mixture of gases, the concentration of each molecular species is commonly given either as moles or as pressure in atmospheres.

Let's use the ammonia equilibrium,

$$N_2(g) + 3H_2(g) \rightleftharpoons 2NH_3(g)$$

where the concentrations are measured as pressures. At 25°C, one study reported the equilibrium pressures of the three gases.

AMMONIA EQUILIBRIUM

Gas	Pressure
NH_3	1.74 atm
H_2	0.06
N_2	0.02

From that experimental data, you can calculate the equilibrium constant for the reaction at 25°C.

$$K = \frac{(p_{NH_3})^2}{(p_{N_2})(p_{H_2})^3} = \frac{(1.74)^2}{(0.02)(0.06)^3} = 7 \times 10^5$$

Equilibrium constants have been determined for many reactions over a wide range of temperatures. One obvious use is to calculate the concentrations of the various substances at equilibrium. In the ammonia example, if you had been given the equilibrium constant K and the pressures of ammonia and hydrogen, you could have calculated the pressure of nitrogen. (Why don't you attempt this simple calculation now?)

Another important use of equilibrium constants is to predict the initial direction of a reaction. Most commonly, the original concentrations of substances are unstable with respect to equilibrium; this is necessarily the case if any substance is entirely absent. The reaction will proceed in one direction until equilibrium is attained. Referring to the ammonia example, if initially

$$\frac{(p_{NH_3})^2}{(p_{N_2})(p_{H_2})^3} > 7 \times 10^5$$

then ammonia decomposes to nitrogen and hydrogen with decreasing p_{NH_3} and increasing p_{N_2} and p_{H_2} until the equilibrium pressures are reached. In the converse case, where initially

$$\frac{(p_{NH_3})^2}{(p_{N_2})(p_{H_2})^3} < 7 \times 10^5$$

nitrogen and hydrogen will combine to form ammonia with decreasing p_{N_2} and p_{H_2} and increasing p_{NH_3} until the values satisfy the equilibrium constant. Notice that only equilibrium concentrations are stable and unchanging.

Although equilibrium calculations involve the concentrations of solutions and gases, the values for pure liquids and solids are virtually constant and so are usually incorporated into the equilibrium constant. As an illustration of this point, the reaction

$$CO(g) + H_2O(l) \rightleftharpoons CO_2(g) + H_2(g)$$

contains a liquid (H_2O) that has a fixed composition. The equilibrium may be expressed by either of the two equations:

$$K_c = \frac{[CO_2][H_2]}{[CO]} \quad \text{or} \quad K_p = \frac{(p_{CO_2})(p_{H_2})}{(p_{CO})}$$

where the concentration of H_2O does not appear because it is included in the equilibrium constant. As was mentioned earlier, there are different equilibrium constants depending on whether the concentrations are molar (K_c) or pressure (K_p).

As one more example of the nonappearance of pure phases in the equilibrium equation, let's examine the reaction of lithium bromide with hydrogen:

$$LiBr(s) + H_2(g) \rightleftharpoons LiH(s) + HBr(g)$$

The equilibrium equation is

$$K = \frac{[HBr]}{[H_2]} = \frac{p_{HBr}}{p_{H_2}}$$

where the values of the two solids of constant composition (lithium bromide and lithium hydride) are in the reported value for K. Although pure liquids and solids do not appear in the equilibrium equation as commonly written, they must be present as real substances for the opposing reactions to be at equilibrium.

The earlier calculations for both acid dissociations and solubility products are special applications of finding concentrations from equilibrium constants.

Problem 40. At 100°C, the halogen reaction

$$Br_2(g) + Cl_2(g) \rightleftharpoons 2BrCl(g)$$

has an equilibrium constant of 0.15 when the concentrations are expressed in either moles or atmospheres. If a reaction vessel is filled with 100 grams of Br_2 and 50 grams each of Cl_2 and BrCl, do an equilibrium calculation to predict the direction of the initial reaction.

Problem 41. At 700°C, the reaction

$$2SO_2(g) + O_2(g) \rightleftharpoons 2SO_3(g)$$

has an equilibrium constant $K_p = 3.76$ when the concentrations of the gases are expressed in atmospheres. If $p_{SO_2} = 0.2$ atm and $p_{O_2} = 0.5$ atm, what is the pressure of SO_3? Assume that the gases are at equilibrium.

Le Chatelier's Principle

A valuable guide is available to assist you in estimating how chemical equilibrium will shift in response to changes in the conditions of the reaction, such as a modification of temperature or pressure. The French chemist Henri Le Chatelier realized in 1884 that if a chemical system at equilibrium is disturbed, the system will adjust itself to minimize the effect of the disturbance. This qualitative reasoning tool is always cited as **Le Chatelier's Principle.**

Let's begin by discussing how an equilibrium system adjusts to a change in the concentration of any substance. At equilibrium, the concentrations of all substances are fixed, and their ratio yields the equilibrium constant. Le Chatelier's Principle tells you that changing the concentration of a substance causes the system to adjust so as to minimize the change in that particular substance. The combustion of methane provides an illustration:

$$CH_4(g) + O_2(g) \rightleftharpoons CO_2(g) + 2H_2O(g)$$

If the four gases in the reaction were at equilibrium, and you then increased the carbon dioxide concentrations, some CO_2 would combine with H_2O to produce CH_4 and O_2 and thereby minimize the increase in CO_2. Alternatively, if you decrease the water vapor concentration by some dehydrating mechanism, some CH_4 would react with O_2 to produce CO_2 and H_2O and thereby minimize any decrease in H_2O. Notice how the concentrations of all constituents shift to counteract the imposed change in a single substance. Of course, this shift does not affect the value of the equilibrium constant.

Now, let's see how a change in pressure would affect the equilibrium reaction

$$N_2(g) + 3H_2(g) \rightleftharpoons 2NH_3(g)$$
$$\text{4 volumes} \qquad \text{2 volumes}$$

which has an equilibrium constant at standard temperature and pressure that is calculated as

$$\frac{(p_{NH_3})^2}{(p_{N_2})(p_{H_2})^3} = K$$

An increase in pressure will, according to Le Chatelier, cause the equilibrium to shift so as to minimize the pressure increase. Because the equilibrium reaction has more relative volumes on the left side, the pressure increase would be minimized by some N_2 and H_2 (total of 4 volumes) combining to NH_3 (2 volumes). Although the relative pressures of the gases have changed, the equilibrium constant still equals K. Conversely, a decrease in pressure would be minimized by the dissociation of some NH_3 (2 volumes) to N_2 and H_2 (4 volumes).

The only reactions that are significantly affected by pressure are those involving gases in which the stoichiometric coefficients of gas molecules add to different values on the two sides of the reaction. Pressure, therefore, would not affect the equilibrium of

$$N_2(g) + O_2(g) \rightleftharpoons 2NO(g)$$

which has 2 volumes on each side. But pressure would affect the equilibrium for

$$2CO(g) + O_2(g) \rightleftharpoons 2CO_2(g)$$

which has 3 volumes on the left and only 2 on the right. In this latter example, an increase of pressure induces the forward reaction, and a decrease in pressure causes the backward reaction. Notice that the effect of varying pressure is to cause the concentrations of the various gases to shift, without any change in the equilibrium constant.

A change in temperature, however, does force a change in the equilibrium constant. Most chemical reactions exchange heat with the surroundings. A reaction that gives off heat is classified as **exothermic**, while a reaction that requires the input of heat is said to be **endothermic**. A simple example of an endothermic reaction is the vaporization of water:

$$H_2O(l) \longrightarrow H_2O(g)$$

which absorbs 10.5 kilocalories per mole. The converse condensation reaction

$$H_2O(g) \longrightarrow H_2O(l)$$

is exothermic because it releases 10.5 kilocalories per mole. A change of state is not required for heat to be involved in a reaction. The combustion of methane

$$CH_4(g) + 2O_2(g) \longrightarrow CO_2(g) + 2H_2O(g)$$

involves only gases, yet this exothermic reaction releases almost 192 kilocalories for every mole of CH_4 that is oxidized. Generally, oxidation reactions are exothermic.

THERMAL CLASSES OF REACTIONS

Role of heat	Description
Reactants + Heat $\longrightarrow$ Products	endothermic
Reactants $\longrightarrow$ Products + Heat	exothermic

In a system at chemical equilibrium, there are always two opposing reactions, one endothermic and the other exothermic.

You can now consider how a change in temperature affects chemical equilibrium. In accordance with Le Chatelier's Principle, the equilibrium constant changes so as to minimize the change in temperature. Beginning with endothermic reactions, an increase in temperature could be minimized by utilizing some of the heat to convert reactants to products, shifting the equilibrium to the right side of the reaction. For exothermic reactions, an increase in temperature could be minimized by using some of the heat to convert "products" into "reactants" and shifting the equilibrium toward the left side.

For a chemical system at equilibrium, an increase in temperature favors the endothermic reaction, while a decrease in temperature favors the exothermic reaction. The equilibrium constant adjusts so that the change in temperature is minimized.

The equilibrium reaction written as

$$CH_4(g) + O_2(g) \rightleftharpoons CO_2(g) + 2H_2O(g)$$

is exothermic when proceeding to the right and endothermic when proceeding to the left. Its equilibrium constant, given by

$$\frac{[CO_2][H_2O]^2}{[CH_4][O_2]} = K$$

must decrease if the temperature increases. Conversely, a lowering of the temperature will cause K to increase.

Please realize that the effect of temperature on the equilibrium constant depends on which of the two opposing reactions is exothermic and which is endothermic. You must have information on the heat of a reaction before you can apply Le Chatelier's Principle to judge how temperature alters the equilibrium.

The next two practice problems refer to the following reaction, which is endothermic in the forward direction.

$$N_2O_4(g) \rightleftharpoons 2NO_2(g)$$

Problem 42. How would an increase in total confining pressure affect the masses of the two nitrogen oxides?

Problem 43. How would an increase in temperature affect the masses of the two nitrogen oxides?

Enthalpy

The experimental discovery that almost all chemical reactions either absorb or release heat led to the idea that all substances contain heat. Consequently, the heat of a reaction is the difference in the heat contents of the various substances:

$$\Delta H = H_{products} - H_{reactants}$$

Throughout the remainder of this book, the Greek letter delta (Δ) will be used to symbolize change. Chemists use the term **enthalpy** for the heat content of a substance or the heat of a reaction, so the H in the equation above means enthalpy. The equation states that the change in enthalpy during a reaction equals the enthalpy of the products minus the enthalpy of the reactants. You can consider enthalpy to be chemical energy that is commonly manifested as heat.

Let's use the decomposition of ammonium nitrate as an example of an enthalpy calculation. The reaction is

$$NH_4NO_3(s) \longrightarrow N_2O(g) + 2H_2O(g)$$

and the enthalpies of the three compounds are given in the next chart.

DATA FOR THE CALCULATION	
Compound	Enthalpy (kcal/mole)
$NH_4NO_3(s)$	−87.4
$N_2O(g)$	19.5
$H_2O(g)$	−57.8

Notice that the enthalpies can be either positive or negative. In general, compounds that released heat when they were formed have a negative enthalpy, and unstable substances that required heat for their formation have a positive enthalpy. The enthalpy of the decomposition reaction would be calculated:

$$\begin{array}{ccc} N_2O & H_2O & NH_4NO_3 \end{array}$$

$$\Delta H = \underbrace{[19.5 + 2(-57.8)]}_{\text{products}} - \underbrace{[-87.4]}_{\text{reactant}} = -8.7\,\text{kcal}$$

Observe the doubling of the enthalpy of H_2O (−57.8) because that compound has a stoichiometric coefficient of 2 in the reaction. The overall enthalpy of the reaction is −8.7 kilocalories, which means that the decomposition of 1 mole of ammonium nitrate releases 8.7 kcal of heat that had been stored within that compound. The release of heat means that this is an exothermic reaction. The sign of the enthalpy of the reaction reveals the direction of heat flow.

ENTHALPY AND HEAT FLOW

Enthalpy of reaction	Type of reaction	Heat
Negative	exothermic	released
Positive	endothermic	absorbed

If you were to reverse the previous reaction to describe the formation of ammonium nitrate,

$$N_2O(g) + 2H_2O(g) \longrightarrow NH_4NO_3(s)$$

then the sign of the enthalpy of the reaction would be reversed:

$$\Delta H = +8.7 \text{ kcal/mole}$$

This particular reaction is, therefore, endothermic. It would require the addition of 8.7 kcal of energy in order to cause the nitrous oxide and water vapor to react to form one mole of ammonium nitrate.

The calculations on the ammonium nitrate reaction demonstrate the immense value of tables that list the enthalpies for various substances. The values at 25°C and 1 atm are called standard enthalpies. For elements, the standard enthalpy is defined as zero. For compounds, the values are called standard enthalpies of formation because they are considered to be formed from elements in their standard state.

The following chart gives a few values that will be used in subsequent examples and problems. The symbol for standard enthalpies of formation is ΔH_f^o, where the superscript denotes standard and the subscript denotes formation. Look up both elemental sulfur and nitrogen to satisfy yourself that the standard enthalpies for elements are 0. Then find the pairs of values for H_2O and CCl_4 (carbon tetrachloride) to learn that the enthalpy depends on the state of matter.

STANDARD ENTHALPIES OF FORMATION
(kcal/mole)

	ΔH_f^o		ΔH_f^o
$Al(s)$	0.0	$MgCl_2(s)$	−153.4
$AlCl_3(s)$	−167.4	$MgO(s)$	−143.8
$Al_2O_3(s)$	−399.1	$Mg(OH)_2(s)$	−221.0
$CCl_4(g)$	−25.5	$N_2(g)$	0.0
$CCl_4(l)$	−33.3	$NH_3(g)$	−11.0
$CO(g)$	−26.4	$NO(g)$	21.6
$CO_2(g)$	−94.0	$N_2O(g)$	19.5
$CaF_2(s)$	−291.5	$O_2(g)$	0.0
$CaO(s)$	−151.9	$O_3(g)$	34.0
$Ca(OH)_2(s)$	−235.8	$S(s)$	0.0
$HCl(g)$	−22.1	$SO_2(g)$	−71.0
$H_2O(g)$	−57.8	$SO_3(g)$	−94.5
$H_2O(l)$	−68.3	$ZnCl_2(s)$	−99.2
$H_2S(g)$	−4.8	$ZnO(s)$	−83.2

Let's use the data in the chart to calculate the enthalpy of the following reaction:

$$2H_2S(g) + 3O_2(g) \longrightarrow 2H_2O(l) + 2SO_2(g)$$

The difference in the heats of formation of the products is given by:

$$\begin{array}{cccc} H_2O & SO_2 & H_2S & O_2 \end{array}$$
$$\Delta H = \underbrace{[2(-68.3) + 2(-71.0)]}_{\text{products}} - \underbrace{[2(-4.8) + 3(0)]}_{\text{reactants}} = -269.0 \text{ kcal}$$

The enthalpy of the reaction is −269 kilocalories, meaning that the oxidation of 2 moles of hydrogen sulfide yields 269 kcal of heat. This reaction is exothermic.

Enthalpy calculations give accurate information on the heat involved in a chemical reaction. They also qualitatively suggest whether a spontaneous reaction is likely to occur. A proposed reaction is most likely to go forward if it is highly exothermic, while an endothermic reaction is unlikely to proceed without the application of external energy. Using this criterion, you can predict that hydrogen sulfide will not persist in the atmosphere because the oxidation reaction that destroys H_2S is highly exothermic. It must be emphasized, however, that the enthalpy of a reaction is not a sure indicator of reactivity, but only a useful suggestion. The following sections discuss other thermodynamic quantities that are needed to determine whether a projected reaction will proceed.

Use the chart Standard Enthalpies of Formation for the next two problems.

Problem 44. Calculate the enthalpy change for the following reaction and classify it as exothermic or endothermic.

$$MgCl_2(s) + H_2O(l) \longrightarrow MgO(s) + 2HCl(g)$$

Problem 45. Calculate the quantity of heat released when 100 grams of calcium oxide react with water.

Energy and Entropy

Thermodynamics is a large body of powerful calculations based on a few fundamental principles. Let's begin by reviewing the two main laws of the field.

The first law of thermodynamics asserts that energy is conserved during any process. The three major forms of energy for chemical purposes are the internal energy of each substance, the external work due to changes in pressure or volume, and the exchange of heat with the surroundings.

The internal energy is sometimes called chemical energy because it is the consequence of all the motions of particles and forces between particles: molecules, atoms, nucleons, and electrons. Particularly important is the energy stored within each chemical bond.

The first law informs us that, in the absence of heat flow, any external work is precisely offset by the opposite change in internal energy. Expansion against a confining pressure reduces the internal energy, whereas external compression of the system increases the internal energy.

The first law of thermodynamics also tells us that if pressure and volume are held constant, any heat flow is counterbalanced by a change in internal energy. An exothermic reaction releasing heat to the surroundings, therefore, is accompanied by a decrease in internal energy, while an endothermic reaction has a concomitant increase in internal energy.

The second law of thermodynamics involves **entropy**, which for our purposes is a statistical measure of the degree of disorder in a chemical system. As an illustration, compare the arrangements of Na^+ and Cl^- ions in both solid and liquid sodium chloride.

Solid sodium chloride has a crystalline structure in which the cations and anions alternate in a repeating pattern.

CRYSTAL STRUCTURE OF SODIUM CHLORIDE

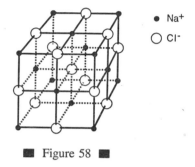

- ● Na⁺
- ○ Cl⁻

■ Figure 58 ■

If you examine the NaCl crystal structure closely, you will see that each Na⁺ is surrounded by 6 Cl⁻, and each Cl⁻ is surrounded by 6 Na⁺. The regular, repetitive structure has a high degree of order and low entropy because entropy varies inversely with order.

When solid NaCl is heated to 801°C it melts, and then the ions are no longer fixed in a simple geometric pattern. They will move relative to each other, subject only to the constraint of electrostatic attraction and repulsion. Each Na⁺, therefore, will be adjacent to as many Cl⁻ anions as possible, and each Cl⁻ will tend to be surrounded by Na⁺ cations. No longer would each ion be surrounded by precisely 6 ions of the opposite charge. This looser arrangement of ions in molten sodium chloride shows that the liquid state displays less order than the solid state, so it is of higher entropy.

A gas has even greater disorder than a liquid because its constituent molecules, atoms, or ions are no longer constrained to be adjacent to each other. Each gas particle moves more or less independently of the other particles. This state is one of minimum order and maximum entropy.

The second law of thermodynamics states that changes within an isolated system are toward higher entropy. This law gives a preferred direction to all processes, including chemical reactions.

Chemists have found it possible to assign a numerical quantity for the entropy of each substance. When measured at 25°C and 1 atm, these are called standard entropies. The next chart lists 12 such values, symbolized by S^o where the superscript denotes the standard state.

STANDARD ENTROPIES
$$\left(\frac{\text{calories}}{\text{deg-mole}} \right)$$

	S^o		S^o
$CO(g)$	47.3	$N_2(g)$	45.8
$CO_2(g)$	51.1	$NO_2(g)$	57.5
$H_2O(l)$	16.7	$N_2O_4(g)$	72.7
$H_2O(g)$	45.1	$O_2(g)$	49.0
$Li(s)$	6.7	$S(s)$	7.6
$Li(g)$	33.1	$SO_2(g)$	59.4

Notice at the top of the chart that the units for entropy require you to multiply each value by the temperature (°K) in order to obtain units of energy.

The point that solids tend to have low entropies and gases usually have high entropies has already been made. An examination of the values in the chart should convince you that this is indeed a valid generalization. Compare the pairs of values for the two states of H_2O and also the two states of lithium.

The symbol for entropy is S, so a change in entropy is shown as ΔS. The values in the chart allow calculations of the entropy change when water evaporates at 25°C.

$$H_2O(l) \longrightarrow H_2O(g)$$

The entropy of reaction is the difference in the values for the products and reactants:

$$\Delta S = S_{products} - S_{reactants} = 45.1 - 16.7 = 28.4 \ cal/deg$$

This positive entropy change means that there is greater disorder in the products (H_2O gas) than the reactants (H_2O liquid). If enthalpy is neglected and only entropy considered, the second law tells us that the H_2O reaction should proceed to the right, toward a condition of higher entropy.

As another example, the entropy change associated with the reaction

$$2CO(g) + O_2(g) \longrightarrow 2CO_2(g)$$

could be calculated from data in the chart of standard entropies:

$$\begin{array}{ccc} CO_2 & CO & O_2 \end{array}$$
$$\Delta S = [2(51.1)] - [2(47.3) + (49.0)] = -41.4 \ cal/deg$$
$$\text{product} \qquad \text{reactants}$$

This negative entropy of reaction would tend to inhibit this reaction from proceeding.

The entropy of reaction by itself, however, is not sufficient to predict the direction of a reaction. At 25°C, you know that $H_2O(l)$ is the stable phase, not $H_2O(g)$. Moreover, the second reaction

$$2CO(g) + O_2(g) \longrightarrow 2CO_2(g)$$

proceeds forward vigorously despite the negative entropy of reaction. You must consider both the enthalpy of reaction and the entropy of reaction in order to determine the direction of a chemical reaction with certainty.

Gibbs Free Energy

The American physicist Josiah Gibbs introduced (ca. 1875) a thermodynamic quantity combining enthalpy and entropy into a single value called **free energy** (or Gibbs free energy). In honor of its inventor, it is usually symbolized as G. The definition of free energy is

$$\underset{\text{free energy}}{G} = \underset{\text{enthalpy}}{H} - \underset{\text{entropy}}{TS}$$

where T is the temperature in degrees Kelvin; entropies must be multiplied by temperature to get units of energy.

For studying chemical reactions, the relationship involves changes in the three thermodynamic quantities:

$$\underset{\substack{\text{free energy}\\\text{of}\\\text{reaction}}}{\Delta G} = \underset{\substack{\text{enthalpy}\\\text{of}\\\text{reaction}}}{\Delta H} - \underset{\substack{\text{entropy}\\\text{of}\\\text{reaction}}}{T\Delta S}$$

A cardinal thermodynamic principle is that systems change toward minimum free energy. The sign of ΔG permits prediction of the behavior of a proposed chemical reaction with certainty.

SIGN OF GIBBS FREE ENERGY

ΔG	Reaction behavior
Negative	proceeds spontaneously
Zero	is at equilibrium
Positive	will not proceed

Although the free energy of a reaction can be calculated from the preceding equation if ΔH and ΔS for the reaction are known, it is much more common to use this alternative equation:

$$\Delta G_{reaction} = \Delta G_{products} - \Delta G_{reactants}$$

where the values on the right are free energies of formation for each substance. For the standard state of 1 atm and 25°C, these values are called standard free energies of formation. As was the case with enthalpies of formation, the standard free energy of formation of an element is defined as zero. The value for each compound is the change in free energy associated with combining the elements to make 1 mole of the compound. In the following chart, the symbol ΔG_f^o has a superscript denoting standard and a subscript denoting formation.

STANDARD FREE ENERGIES
OF FORMATION
(kcal/mole)

	ΔG_f^o		ΔG_f^o
$CO(g)$	−32.8	$NO(g)$	20.7
$CO_2(g)$	−94.3	$NO_2(g)$	12.4
$H_2(g)$	0.0	$O_2(g)$	0.0
$H_2O(g)$	−54.6	$S(s)$	0.0
$H_2O(l)$	−56.6	$SO_2(g)$	−71.8
$H_2S(g)$	−7.9	$SO_3(g)$	−88.5

Given a listing of free energies of formation, the free energy change of a chemical reaction may be calculated in the same manner as you evaluated enthalpies of reaction and entropies of reaction. For the reaction that was discussed earlier,

$$2CO(g) + O_2(g) \rightleftharpoons 2CO_2(g)$$

the free energy of reaction is calculated

$$\begin{array}{cc} CO_2 & CO \quad O_2 \\ \Delta G = [2(-94.3)] - [2(-32.8) + (0)] = -123.0 \text{ kcal} \\ \text{product} & \text{reactants} \end{array}$$

The negative free energy of reaction means that the forward reaction will occur spontaneously. Remember, systems change toward a decrease in free energy.

As another example of this important technique, let's consider the reaction involving oxides of both carbon and sulfur:

$$CO_2(g) + SO_2(g) \rightleftharpoons CO(g) + SO_3(g)$$

From the free energies of formation given in the chart on page 149, you can calculate the free energy of the reaction:

$$\begin{array}{cc} CO \quad SO_3 & CO_2 \quad SO_2 \\ \Delta G = [(-32.8) + (-88.5)] - [(-94.3) + (-71.8)] = +44.8 \text{ kcal} \\ \text{products} & \text{reactants} \end{array}$$

This positive value for the free energy of reaction means that the reaction will not proceed to the right. Instead, the reverse reaction is favored, with CO and SO_3 reacting to yield CO_2 and SO_2. Of course, this reaction can take place only if both carbon monoxide and sulfur trioxide are present initially. If you started with only carbon dioxide and sulfur dioxide, no reaction could occur.

The free energy calculations for both reactions

$$2CO + O_2 \rightleftharpoons 2CO_2$$

and

$$CO_2 + SO_2 \rightleftharpoons CO + SO_3$$

are valid only at 25°C because standard free energies of formation were utilized to calculate the free energy changes of the two reactions.

However, if you know both the standard enthalpies of formation and standard entropies for every substance in a reaction, then you can estimate the free energy of reaction at other temperatures by using the equation

$$\Delta G = \Delta H - T\Delta S$$

which, you'll recall, is Gibbs' definition for free energy changes.

Let's use the preceding equation to estimate the free energy change at 400°C for the reaction

$$2CO(g) + O_2(g) \rightleftharpoons 2CO_2(g)$$

You need tabulated values of standard enthalpies of formation and standard entropies for the three gases, as shown in the following chart.

	VALUES AT 25°C	
Gas	ΔH_f^o (kcal/mole)	S^o (cal/deg-mole)
$CO(g)$	−26.4	47.3
$CO_2(g)$	−94.0	51.1
$O_2(g)$	0.0	49.0

Using those values, you should be able to calculate the changes for the reaction at 25°C and obtain

$$\Delta H = -135.2 \text{ kcal}$$
$$\Delta S = -41.4 \text{ cal/deg}$$

You should perform both calculations right now to make sure that you understand how they were done. Notice that the energy units for the values differ by a factor of 1,000; in the calculations that follow, ΔS is divided by 1,000 to make the terms comparable.

Let's use those values now to calculate the free energy of reaction from the equation:

$$\Delta G = \Delta H - T\Delta S$$

At 25°C, which equals 298°K,

$$\Delta G = (-135.2) - (298)(-0.0414) = -122.9 \text{ kcal}$$

At 400°C, which equals 673°K,

$$\Delta G = (-135.2) - (673)(-0.0414) = -107.3 \text{ kcal}$$

The two results for the free energy of reaction differ because the equilibrium of the chemical reaction changes with temperature. The first free energy change (at 25°C) is accurate because the tabulated values of enthalpies of formation and entropies were valid at that standard temperature. The second free energy change (at 400°C) is only a convenient estimate, based on the assumption that ΔH and ΔS are constants. Actually, both those values do change slightly with temperature.

This final chapter has only begun to introduce you to the many thermodynamic calculations that assist chemists in understanding reactions.

THREE THERMODYNAMIC QUANTITIES

Quantity	Symbol	Measures	Units
Enthalpy	H	heat	energy
Entropy	S	disorder	energy/deg
Free energy	G	reactivity	energy

Don't forget that both the calorie and the joule are units of energy in published charts, so you will often have to convert to obtain the unit that you want.

Problem 46. Use the chart of standard free energies of formation (page 149) to determine the free energy change in the proposed reaction

$$NO(g) + H_2O(l) \longrightarrow NO_2(g) + H_2(g)$$

and report whether the reaction would proceed.

Problem 47. Use the standard values in the chart on the next page to compare the free energy change of the reaction

$$N_2O_4(g) \rightleftharpoons 2NO_2(g)$$

at both 25°C and 100°C.

DATA FOR THE CALCULATION

	ΔH_f° (kcal/mole)	S° (cal/deg-mole)
$N_2O_4(g)$	2.309	72.73
$NO_2(g)$	8.088	57.47

Glossary of Chemical Terms

acid a compound that yields H^+ ions in solution or a solution with the concentration of H^+ exceeding OH^-

acid ionization constant the equilibrium constant describing the degree of dissociation of an acid

actinides the row of elements below the periodic table, from thorium to lawrencium

alkali synonym for base

alkali metals the column of elements from lithium to francium

alkaline earths the column of elements from beryllium to radium

alkane a hydrocarbon without a double bond, triple bond, or ring structure

alkene a hydrocarbon with one or more double bonds and no triple bond

alkyne a hydrocarbon with one or more triple bonds

alpha particle a cluster of 2 protons and 2 neutrons emitted from a nucleus in one type of radioactivity

anion an atom or molecule with a negative charge

anode the negative electrode at which oxidation occurs

aqueous refers to a solution with water as solvent

aromatic refers to an organic compound with a benzene-like ring

atom the smallest amount of an element; a nucleus surrounded by electrons

atomic number the number of protons in the nucleus of the chemical element

atomic weight the weight in grams of one mole of the chemical element; approximately the number of protons and neutrons in the nucleus

Avogadro's Law equal volumes of gases contain the same number of molecules

Avogadro's Number 6.02×10^{23}, the number of molecules in 1 mole of a substance

base a compound that yields OH⁻ ions in solution or a solution with the concentration of OH⁻ exceeding H⁺

beta particle an electron emitted from a nucleus in one type of radio-activity

boiling point the temperature at which a liquid changes to a gas

boiling point elevation. an increase in the boiling point of a solution, proportional to the concentration of solute

Boyle's Law the volume of a gas varies inversely with pressure

calorie a unit of energy, equal to 4.184 joules

catalyst a substance that accelerates a chemical reaction without itself being a reactant

cathode the positive electrode at which reduction occurs

cation an atom or molecule with a positive charge

Charles' Law the volume of a gas varies directly with temperature

compound a substance formed by the chemical combination of two or more elements

concentration the relative abundance of a solute in a solution

congeners elements with similar properties, found in one column of the periodic table

conjugate an acid and base that are related by removing or adding a single hydrogen ion

covalent bond atoms linked together by sharing valence electrons

critical point a point in a phase diagram where the liquid and gas states cease to be distinct

crystalline the regular, geometric arrangement of atoms in a solid

decomposition a chemical reaction in which a compound is broken down into simpler compounds or elements

dissociation the separation of a solute into constituent ions

electric cell a device that uses a chemical reaction to produce an electric current

electrode a conducting substance that connects an electrolyte to an external circuit

electrolysis the decomposition of a substance by an electric current

electrolyte an ionic substance that has high electrical conductivity

electromotive force the electrical potential produced by a chemical reaction

electron a light subatomic particle with negative charge; found in orbitals surrounding an atomic nucleus

electronegativity a number describing the attraction of an element for electrons in a chemical bond

element a substance that cannot be decomposed; each chemical element is characterized by the number of protons in the nucleus

EMF see *electromotive force*

endothermic refers to a reaction that requires heat

energy the concept of motion or heat

enthalpy the thermodynamic quantity measuring the heat of a substance

entropy the thermodynamic quantity measuring the disorder of a substance

equilibrium a balanced condition resulting from two opposing reactions

equilibrium constant the ratio of concentrations of products to reactants for a reaction at chemical equilibrium

exothermic refers to a reaction that releases heat

faraday a unit of electric charge equal to that on 1 mole of electrons

Faraday's Laws two laws of electrolysis relating the amount of substance to the quantity of electric charge

fluid a liquid or gas

free energy the thermodynamic quantity measuring the tendency of a reaction to proceed; also called Gibbs free energy

freezing point the temperature at which a liquid changes to a solid

freezing point depression the decrease in freezing point of a solution, proportional to the concentration of solute

fusion melting

gas constant R equals 0.082 liter-atmospheres per mole-degree

gram formula weight an amount of a substance equal in grams to the sum of the atomic weights

ground state the electronic configuration of lowest energy for an atom

group a column of elements in the periodic table

half-reaction an oxidation or reduction reaction with free electrons as a product or reactant

halogens the column of elements from fluorine to astatine

heat capacity the amount of energy needed to raise the temperature of a substance by 1°

hydrocarbon an organic compound containing only carbon and hydrogen

hydrogen bond a weak, secondary bond between a negative ion and a hydrogen atom that has a strong primary polar bond to another atom

hydroxyl refers to the OH⁻ ion

inert gases the column of elements from helium to radon; also called noble gases

ion an atom with an electric charge due to gain or loss of electrons

ionic bond atoms linked together by the attraction of unlike charges

ionization adding or subtracting electrons from an atom; alternatively, the dissociation of a solute into ions

isoelectronic refers to several dissimilar atoms or ions with identical electronic configurations

isomers several molecules with the same composition but different structures

isotope a variety of an element characterized by a specific number of neutrons in the nucleus

joule a unit of energy equal to 0.239 calorie

lanthanides the row of elements beneath the periodic table, from cerium to lutetium; also called rare earths

Le Chatelier's Principle a system that is disturbed adjusts so as to minimize the disturbance

litmus an indicator that turns red in acid and blue in alkaline solution

melting point the temperature at which a solid changes to a liquid

metallic bond atoms linked together by the migration of electrons from atom to atom

metals the elements in the middle and left parts of the periodic table, except for hydrogen

molality the number of moles of solute in 1 kilogram of solvent

molarity the number of moles of solute in 1 liter of solution

mole an amount of a substance equal in grams to the sum of the atomic weights

molecular formula describes the ratio of elements in a molecule

molecule a group of atoms linked together by covalent bonds

neutralization the chemical reaction of an acid and base to yield a salt and water

neutron a heavy subatomic particle with zero charge; found in an atomic nucleus

noble gases the column of elements from helium to radon; also called inert gases

nonmetals the elements in the upper right part of the periodic table, and also hydrogen

nucleon a proton or neutron found in an atomic nucleus

nucleus the core of an atom, containing protons and neutrons

orbital a classification of the energy level occupied by up to 2 electrons

organic refers to compounds based on carbon

oxidation a reaction involving the loss of electrons by an element

oxidation number a signed integer representing the real or hypothetical charge on an atom

oxide a compound of oxygen and another element

periodic table display of the elements in order of atomic number with similar elements falling into columns

pH a number describing the concentration of hydrogen ions in a solution

phase a substance with uniform composition and definite physical state

polar bond atomic linkage with both ionic and covalent characteristics

polyprotic refers to an acid with several hydrogens that can dissociate

precipitate a solid that separates from solution

product a substance on the right side of a chemical reaction

proton a heavy subatomic particle with a positive charge; found in an atomic nucleus

radioactivity the emission of subatomic particles from a nucleus

rare earths the elements from cerium to lutetium; lanthanides

reactant a substance on the left side of a chemical reaction

redox refers to a reaction with simultaneous reduction and oxidation

reduction a reaction involving the gain of electrons by an element

salt a solid compound composed of both metallic and nonmetallic elements

saturated describes a solution that has as much solute as possible

shell a set of electron orbitals with the same principal quantum number

solubility the upper limit to the concentration of a solute

solubility product the constant obtained by multiplying the ion concentrations in a saturated solution

solute the substance that is dissolved in a solution

solvent the host substance of dominant abundance in a solution

standard temperature and pressure 0°C and 1 atmosphere

states of matter solid, liquid, and gas

stoichiometric refers to compounds or reactions in which the components are in fixed, whole-number ratios

STP see *standard temperature and pressure*

strong electrolyte an acid, base, or salt that dissociates almost completely to ions in aqueous solution

structural formula depicts the bonding of atoms in a molecule

sublimation the transformation of a solid directly to a gas without an intervening liquid state

subshell a set of electron orbitals with the same principal and second quantum numbers

transition metals the three rows of elements in the middle of the periodic table, from scandium to mercury

triple point a point in a phase diagram where the three states of matter are in equilibrium

valence a signed integer describing the combining power of an atom as a real or hypothetical charge

valence electrons the outermost shell of electrons

weak electrolyte an acid, base, or salt that dissociates only slightly to ions in an aqueous solution

Answers to Problems

1. By weight, the compound is 26.19% nitrogen, 7.55% hydrogen, and 66.26% chlorine.

Nitrogren	$1 \times 14.02 =$	$14.01/53.50 = 0.2619$
Hydrogen	$4 \times 1.01 =$	$4.04/53.50 = 0.0755$
Chlorine	$1 \times 35.45 =$	$\underline{35.45}/53.50 = 0.6626$
		53.50 Total

2. The simplest formula is K_2CuF_4.

Potassium	$\dfrac{35.91}{39.10} = 0.918/0.459 = 2$
Copper	$\dfrac{29.19}{63.55} = 0.459/0.459 = 1$
Fluorine	$\dfrac{34.90}{19.00} = 1.837/0.459 = 4$

3. There are 6.37 moles of C_6H_5Br.

   $$1 \text{ mole} = 6(12.01) + 5(1.01) + 1(79.90) = 157.01 \text{ grams}$$

   $$\frac{1000 \text{g } C_6H_5Br}{157.01 \text{ g/mole}} = 6.37 \text{ moles}$$

4. The neon has a mass of 4.5 grams.

   $$\frac{5 \text{ liters Ne}}{22.4 \text{ lit/mole}} = 0.223 \text{ moles} \times 20.18 \text{ g/mole} = 4.5 \text{ grams}$$

5. The reaction uses 2.79 liters of oxygen.

$$1 \text{ mole } CH_4 = 1(12.01) + 4(1.01) = 16.05 \text{ grams}$$

$$\frac{1 \text{ gram}}{16.05 \text{ g/mole}} = 0.0623 \text{ mole} \times 22.4 = 1.395 \text{ liters } CH_4$$

The reaction coefficients denote the relative volumes, so the volume of O_2 is twice that of $CH_4 = 2 \times 1.395 = 2.79$ liters.

6. Nuclei B and C are isotopes of magnesium, the element with atomic number 12. Nuclei A and B both have a mass of approximately 24 atomic mass units because their nucleons sum to 24.

7. Natural silver is 48.15% silver-109.

$$\text{Let } x = \text{fraction } ^{109}Ag \text{ and } (1 - x) = \text{fraction } ^{107}Ag$$
$$108.905(x) + 106.905(1 - x) = 107.868$$
$$2x + 106.905 = 107.868$$
$$x = 0.4815$$

8. The nucleus is radium-226, also written $^{226}_{88}Ra$. The atomic mass decreased by 4, the mass of the alpha particle. The atomic number decreased by 2 because the alpha particle carried off 2 protons.

9. Aluminum has 3 valence electrons, whereas oxygen has 6. Remember that you count columns from the left margin of the periodic table.

10. The Lewis diagram for H_2S is

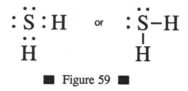

■ Figure 59 ■

11. The electronegativity difference of magnesium and chlorine is 1.8,

$$3.0 - 1.2 = 1.8$$
$$Cl \quad Mg$$

which corresponds to a bond with 52% ionic character and 48% covalent character. Such an intermediate bond is called polar.

12. The three isomers of C_5H_{12} are shown below. The essential feature is the bonding of carbons. In the first molecule, no carbon is bonded to more than 2 carbons, the second molecule has a carbon bonded to 3 carbons, and the third molecule has a carbon bonded to 4 carbons.

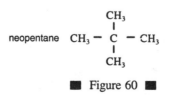

n-pentane $CH_3 - CH_2 - CH_2 - CH_2 - CH_3$

isopentane $CH_3 - CH - CH_2 - CH_3$
                         $|$
                       CH_3

neopentane $CH_3 - C - CH_3$
                       $|$
                     CH_3

■ Figure 60 ■

13. The addition of hydrogen converts acetylene to ethane:

$$C_2H_2 \quad + \quad 2H_2 \quad \longrightarrow \quad C_2H_6$$

acetylene   hydrogen   ethane

Because the number of moles of hydrogen is twice that of acetylene, the reaction requires 200 liters of hydrogen, double that of acetylene.

14. It is an aldehyde with the structural formula

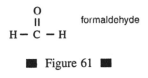

■ Figure 61 ■

15. The minimum pressure for liquid CO_2 is 5.1 atmospheres. At the triple point, $\ln(P) = 1.63$; therefore, $P = e^{1.63} = 5.1$.

16. At $-64°C$, the solid CO_2 sublimates to the gas state. The vertical scale on the phase diagram is logarithmic and $\ln(3) = 1.10$. Follow that value to the right.

17. The total heat needed is 11,910 calories.

$$5.55 \times 40 \times \frac{8.74 \text{ calories}}{\text{deg-mole}} = 1940 \text{ calories}$$

moles   degrees   (warming ice to 0°)

$$5.55 \times \frac{1436 \text{ calories}}{\text{mole}} = 7970 \text{ calories}$$

moles   (heat of fusion)

$$5.55 \times 20 \times \frac{18.02 \text{ calories}}{\text{deg-mole}} = 2000 \text{ calories}$$

moles   degrees   (warming water to 20°)

18. The pressure equals 0.804 atmosphere.

$$611 \text{ mm Hg} \times \frac{1 \text{ atm}}{760 \text{ mm Hg}} = 0.804 \text{ atm}$$

19. The required pressure is 1.654 atmospheres.

$$(1.13 \text{ atm})(732 \text{ ml}) = (x \text{ atm})(500 \text{ ml})$$
$$x = (1.13 \text{ atm}) \frac{732 \text{ ml}}{500 \text{ ml}} = 1.654 \text{ atm}$$

20. The chilled temperature is $-217.63°C$.

$$\frac{V}{T} = \frac{660 \text{ ml}}{293.15°K} = \frac{125 \text{ ml}}{x}$$
$$x = (293.15°K) \frac{125 \text{ ml}}{660 \text{ ml}} = 55.52°K = -217.63°C$$

21. There are 1.5×10^{24} hydrogen atoms.

$$\text{moles } CH_4 = \frac{10 \text{g } CH_4}{16.05 \text{ g/mole}} = 0.623 \text{ mole}$$

$$\text{atoms } H = (6.02 \times 10^{23})(0.623 \text{ mole})(4 \text{ atoms/molecule})$$
$$= 1.5 \times 10^{24}$$

22. The carbon monoxide occupies 28,489 liters.

$$\frac{1000 \text{ grams}}{28.01 \text{ g/mole}} = 35.70 \text{ moles CO}$$

$$V = \frac{nRT}{P} = \frac{(35.70)(0.082)(973.15)}{(0.1)} = 28,489 \text{ liters}$$

23. The ozone molecule has the formula O_3.

$$\frac{1.073 \text{ liters}}{22.4 \text{ lit/mole}} = 0.0479 \text{ mole}$$

$$\frac{2.3 \text{ grams}}{0.0479 \text{ mole}} = 48 \text{ grams/mole (molecular weight)}$$

$$\frac{48 \text{ g/mole ozone molecules}}{16 \text{ g/mole oxygen atoms}} = 3 \text{ atoms per ozone molecule}$$

24. The solution is 0.592 m in glucose.

$$C_6H_{12}O_6 = 6(12.01) + 12(1.01) + 6(16.00) = 180.18 \text{ grams/mole}$$

$$\text{moles glucose} = \frac{80}{180.18} = 0.444$$

$$\text{molality} = \frac{0.444 \text{ mole glucose}}{0.750 \text{ gram solvent}} = 0.592 \text{ m}$$

25. The solution is 0.36 mole fraction alcohol.

$$CH_3OH = 32.05 \text{ grams/mole}$$

$$H_2O = 18.02 \text{ grams/mole}$$

$$\text{moles alcohol} = 100/32.05 = 3.12$$

$$\text{moles water} = 100/18.02 = 5.55$$

$$\text{mole fraction alcohol} = \frac{(3.12)}{(3.12 + 5.55)} = 0.36$$

26. The amount of CuCl is 0.00152 mole. If the powder dissolved completely, the solution would be 0.00152 molar with respect to both Cu^+ and Cl^-.

$$[Cu^+] \ [Cl^-] = (0.00152)^2 = 2.3 \times 10^{-6}$$

Because that product exceeds the solubility product given in the chart as 1.1×10^{-6}, which is the value for a saturated solution, the powder will not completely dissolve.

27. The solubility of aluminum hydroxide is 0.00843 gram per liter. The $Al(OH)_3$ dissociates to 4 ions with the concentration of OH^- being 3 times that of Al^{3+}.

$$[Al^{3+}] \ [OH^-]^3 = K_{sp}$$
$$(x)(3x)^3 = 3.7 \times 10^{-15}$$
$$27x^4 = 3.7 \times 10^{-15}$$
$$x^4 = 1.37 \times 10^{-16}$$
$$x = 1.08 \times 10^{-4} = \text{molarity of } Al(OH)_3$$

$$(1.08 \times 10^{-4} \text{ mole})(78.01 \text{ g/mole}) = 0.00843 \text{ gram}$$

28. The sodium chloride solution boils at 100.873°C.

$$\frac{10 \text{g NaCl}}{58.44 \text{ g/mole}} = 0.1711 \text{ mole}$$

$$\frac{0.1711 \text{ mole}}{0.2 \text{ Kg H}_2\text{O}} = 0.8556 \text{ molal re NaCl}$$

Each formula unit yields 2 ions; so, the total molality of the ionic solution is twice that, or 1.7112 m. The change in boiling point is

$$\Delta T_b = K_b m = (0.51)(1.7112) = 0.873°$$

and that value is added to the 100° boiling point of pure water.

29. The molecular weight of brucine is approximately 394. The chart on page 89 states that pure chloroform freezes at −63.5°.

$$\Delta T_f = (-63.5) - (-64.69) = 1.19°$$

$$m = \frac{\Delta T_f}{K_f} = \frac{1.19}{4.68} = 0.254 \text{ molal}$$

$$\frac{100 \text{ grams}}{0.254 \text{ mole}} = 393.7 \text{ grams/mole}$$

30. The solution is alkaline with pH = 8.34.

$$[H^+] = \frac{K_w}{[OH^-]} = \frac{1 \times 10^{-14}}{2.2 \times 10^{-6}} = 4.55 \times 10^{-9}$$

$$pH = -\log_{10}(4.55 \times 10^{-9}) = 8.34 \text{ (over 7, so alkaline)}$$

31. The solution required 0.0566 mole of acetic acid.

From the pH, $[H^+] = 10^{-3}$ and $[CH_3COO^-]$ must be the same.

$$\frac{[H^+] [CH_3COO^-]}{[CH_3COOH]} = \frac{(10^{-3})(10^{-3})}{x - 10^{-3}} = 1.8 \times 10^{-5} \text{ } (K_a \text{ from table})$$

$$x - 10^{-3} = \frac{10^{-6}}{1.8 \times 10^{-5}} = 5.56 \times 10^{-2}$$

$$x = (5.56 \times 10^{-2}) + (1 \times 10^{-3}) = 5.66 \times 10^{-2}$$

32. The conjugate base for HCO_3^- is the carbonate ion CO_3^{2-}, formed by the donation of a proton. The conjugate acid is carbonic acid H_2CO_3, formed as HCO_3^- accepts a proton.

33. The ratio of $[H_2PO_4^-]$ to $[PO_4^{3-}]$ equals 5.04×10^{11}.

$$\frac{[H_2PO_4^-]}{[HPO_4^{2-}]} = \frac{[H^+]}{6.2 \times 10^{-8}} = \frac{10^{-4}}{6.2 \times 10^{-8}} = 1.61 \times 10^3$$

$$\frac{[HPO_4^{2-}]}{[PO_4^{3-}]} = \frac{[H^+]}{3.2 \times 10^{-13}} = \frac{10^{-4}}{3.2 \times 10^{-13}} = 3.13 \times 10^8$$

$$\frac{[H_2PO_4^-]}{[PO_4^{3-}]} = \frac{[H_2PO_4^-]}{[HPO_4^{2-}]} \frac{[HPO_4^{2-}]}{[PO_4^{3-}]} = (1.61 \times 10^3)(3.13 \times 10^8)$$

$$= 5.04 \times 10^{11}$$

34. Nitrogen has the oxidation number -3 in Mg_3N_2 and $+5$ in HNO_3.

For Mg_3N_2,

$$\begin{array}{cc} Mg & N \\ \end{array}$$
$$3(+2) + 2(x) = 0 \text{ so } x = -3$$

For HNO_3,

$$\begin{array}{ccc} H & N & O \\ \end{array}$$
$$1(+1) + 1(x) + 3(-2) = 0 \quad \text{so } x = +5$$

Notice that the oxidation numbers are multiplied by the number of atoms in each formula unit.

35. Carbon is oxidized and iodine is reduced, so CO is the reducing agent and I_2O_5 is the oxidizing agent.

$$I_2^{(+5)}O_5^{(-2)} + 5C^{(+2)}O^{(-2)} \longrightarrow I_2^{(0)} + 5C^{(+4)}O_2^{(-2)}$$

Each of the 5 carbon atoms loses 2 electrons, and each of the 2 iodine atoms gains 5 electrons.

36. Only manganese and oxygen have variable oxidation numbers.

$$\overset{\text{loss of 3 electrons}}{\overbrace{Mn^{(+7)}O_4^{(-2)} \qquad Mn^{(+4)}O_2^{(-2)} + O_2^{(0)}}}$$

gain of 4 electrons per O_2

$$\overset{\text{loss of 12 electrons}}{\overbrace{4Mn^{(+7)}O_4^{(-2)} \qquad 4Mn^{(+4)}O_2^{(-2)} + 3O_2^{(0)}}}$$

gain of 12 electrons

Using these coefficients and balancing charges,

$$4MnO_4^- + 4H^+ \longrightarrow 4MnO_2 + 3O_2 + H_2O$$

Finish by balancing the number of hydrogens:

$$4MnO_4^- + 4H^+ \longrightarrow 4MnO_2 + 3O_2 + 2H_2O$$

37. The silver is deposited from solution as the iron dissolves.

oxidation $Fe(s) \longrightarrow Fe^{2+}(aq) + 2e^-$

reduction $2Ag^+(aq) + 2e^- \longrightarrow 2Ag(s)$

overall $Fe(s) + 2Ag^+(aq) \longrightarrow 2Ag(s) + Fe^{2+}(aq)$

38. The lithium-fluorine battery yields 5.91 volts.

Half-reaction	Type	Potential
$2Li \longrightarrow 2Li^+ + 2e^-$	oxidation	3.04
$F_2 + 2e^- \longrightarrow 2F^-$	reduction	2.87

39. The electrolysis requires 111.2 faradays of electricity.

$$\frac{1000 \text{ grams Al}}{26.98 \text{ g/mole}} = 37.06 \text{ moles Al}$$

$Al^{(+3)} + 3e^- \longrightarrow Al^{(0)}$ (reduction)

moles of electrons $= 3 \times$ moles $Al = 3 \times 37.06 = 111.2$

40. The initial reaction is $2BrCl \longrightarrow Br_2 + Cl_2$

moles BrCl $=$ 50/115.35 $= 0.433$

moles Br_2 $=$ 100/159.80 $= 0.626$

moles Cl_2 $=$ 50/70.90 $= 0.705$

$$\frac{[BrCl]^2}{[Br_2][Cl_2]} = \frac{(0.433)^2}{(0.626)(0.705)} = 0.425 > 0.15$$

41. The value of p_{SO_3} is 0.274 atmosphere.

$$\frac{(p_{SO_3})^2}{(p_{SO_2})^2(p_{O_2})} = 3.76$$

$$(p_{SO_3})^2 = (3.76)(0.2)^2(0.5) = 0.0752$$

42. The mass of N_2O_4 would increase, and NO_2 would decrease. The volume coefficient of the left side (1) is less than that of the right side (2), so a conversion of NO_2 to N_2O_4 would minimize the increase in pressure.

43. The mass of NO_2 would increase, and N_2O_4 would decrease. Because the forward reaction is endothermic,

$$N_2O_4 + heat \rightleftharpoons 2NO_2$$

the conversion of N_2O_4 to NO_2 would absorb heat and minimize the increase in temperature.

44. The enthalpy of reaction is +33.7 kcal, so the reaction is endothermic.

$$\text{MgO} \qquad \text{HCl} \qquad\qquad \text{MgCl}_2 \qquad \text{H}_2\text{O}$$
$$\Delta H = [(-143.8) + 2(-22.1)] - [(-153.4) + (-68.3)] = +33.7 \text{ kcal}$$

45. The exothermic reaction releases 27.8 kilocalories of heat.

$$\text{CaO}(s) + \text{H}_2\text{O}(l) \longrightarrow \text{Ca(OH)}_2(s)$$

$$\Delta H = [-235.8] - [(-151.9) + (-68.3)] = -15.6 \text{ kcal/mole}$$
$$\text{product} \qquad\qquad \text{reactants}$$

$$\Delta H = (1.783 \text{ moles CaO})(-15.6 \text{ kcal/mole}) = -27.8 \text{ kcal}$$

46. The free energy change is 48.3 kcal; because this is positive, the reaction would not proceed.

$$\text{NO}_2 \quad \text{H}_2 \qquad\quad \text{NO} \qquad \text{H}_2\text{O}$$
$$[(12.4) + (0)] - [(20.7) + (-56.6)] = 48.3$$
$$\text{products} \qquad\qquad \text{reactants}$$

47. The temperature change reverses the reaction direction. From the standard values that were given, you could calculate that

$$\Delta H = 13.867 \text{ kcal}$$
$$\Delta S = 42.21 \text{ cal/deg} = 0.04221 \text{ kcal/deg}$$

and then substitute those into

$$\Delta G = \Delta H - T\Delta S$$

At 25°C = 298°K, the free energy favors N_2O_4:

$$\Delta G = (13.867) - (298)(0.04221) = +1.29 \text{ kcal}$$

At 100°C = 373°K, the free energy favors NO_2:

$$\Delta G = (13.867) - (373)(0.04221) = -1.88 \text{ kcal}$$